KU-747-770

INTERVALS, SCALES AND TEMPERAMENTS

LL. S. LLOYD AND HUGH BOYLE

INTERVALS, SCALES AND TEMPERAMENTS

PART I

ARTICLES BY THE LATE LL. S. LLOYD SELECTED AND EDITED WITH AN INTRODUCTION AND BIBLIOGRAPHY BY HUGH BOYLE.

PART II

DEFINITIONS, EXPERIMENTS, MEASUREMENTS, CALCULATIONS, VARIOUS APPENDICES AND TABLES BY HUGH BOYLE

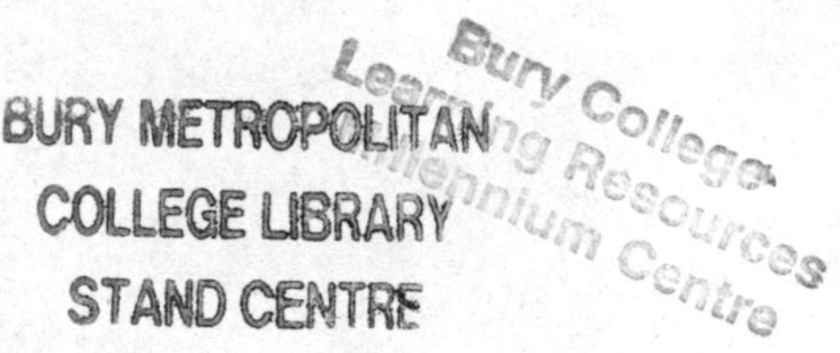

MACDONALD AND JANE'S · LONDON

First published in 1963
This edition published in Great Britain
by Macdonald and Jane's Publishers Limited
Paulton House, 8 Shepherdess Walk,
London N1 7LW

Printed in Great Britain by
REDWOOD BURN LIMITED
Trowbridge & Esher

To Henry Washington
and his Schola Polyphonica
whose beauty of intonation inspired me to make these studies.

BIOGRAPHICAL NOTE

LLOYD, LLEWELYN, S. (SOUTHWORTH) (*b.* Cheadle Hulme, Cheshire, 20 Apr. 1876. *d.* Birmingham, 14 Aug. 1956).

British physicist and writer on acoustics. He was educated at King William's College in the Isle of Man and at Christ's College, Cambridge, where he was 9th Wrangler in 1898 and took a First Class in the Natural Science Tripos in 1899. He became H.M. Inspector of Schools (Secondary Schools branch) in 1905 and in 1917 Assistant Secretary to the Department of Scientific Research. From 1935 to 1943 he was Principal Assistant Secretary. He was created C.B. (Civil) in 1921. From 1946 to 1950 he was chairman of the Committee on Standard Pitch, British Standards Institution.

Lloyd has written numerous articles for scientific and musical journals on the physical aspects of music: sound, acoustics, intonation, temperaments, standard pitch, bells, etc., as well as new articles dealing with such matters for the present edition of this Dictionary which have thus been brought into line with modern ideas and research. His books, *Music and Sound* (Oxford, 1937), *A Musical Slide-Rule* (Oxford, 1938) and *The Musical Ear* (Oxford, 1940) all approach their subjects from the musician's point of view.

The great value of the work done by Lloyd as a physicist on behalf of music lies in the fact that, although a scientist by training, he is also a musician by inclination and education. For him, whenever he deals with a musical subject, art comes first and science takes a secondary place. That is to say, if a scientific explanation does not account for a musical phenomenon to his satisfaction, he concludes that scientific knowledge is incomplete. With the true scientist's instinct he mistrusts all would-be scientific assumptions until he has found them to stand up to the test of the music of great masters. His criticism of 19th-century scientists who dealt with musical matters does not spare the most eminent where he thinks them at fault, but does not apply, among others, to William Pole and Blaikley, nor, in particular, to Helmholtz, for whose work he has a great admiration, but who, he thinks, has been largely misinterpreted in England. E.B.

—from 'Grove's Dictionary of Music', by permission of the publishers, Messrs. Macmillan & Co. Ltd.

Music is, however, an art not of notes but of intervals.

EDWIN EVANS

(from his article on Atonality and Polytonality in Cobbett's *Cyclopedic Survey of Chamber Music*)

Whether or no a particular interval is appropriate to the musical occasion only the ear of the musician can determine.

LL. S. LLOYD

(from the last chapter of *The Musical Ear*)

FOREWORD

BY KENNETH VAN BARTHOLD,
L.R.A.M., LAURÉAT DU CONSERVATOIRE DE PARIS

Mr. van Barthold is a professional pianist, a teacher of the pianoforte at Trinity College of Music, and head of the music department at the City Literary Institute.

WHEN Hugh Boyle first persuaded me to listen to a 'pure' sixth followed by a tempered one, as produced on the piano, the restfulness and beauty of the first was so striking, and the harshness of the second so disturbing, that I was shocked. From here I went on to experiment with the intonation of various intervals in different contexts (both harmonic and melodic) on a calibrated monochord of the kind he describes in this book. The B in the major triad on G can be felt to rise when the Tritone F is sounded below. The D a fifth above G and the A a third above F form a very dubious fifth, if the G and F are perfect fourths on either side of C (see Lloyd's article 'The Perfect Fifth' on page 12).

The point at issue is the fixing of 'pitch'. Mr. Boyle argues that sounds should be pitched according to the dictates of the ear. For this certain notes in the scale must be free to move about a comma, which is perfectly possible on all but keyboard instruments.

What has happened is that the keyboard has come to be the arbiter of intonation. Many a singer is brought up suddenly by a bang on the piano. But piano intonation equivocates; the sounds are impure, many of the overtones lost or damped on purpose, and every interval except the octave out of tune (and many octaves are now 'stretched' for added brilliance). This then is the arbiter we use; more, it is for many students their first contact with the quality and classification of tonal intervals and harmonies. The dangers to the sensitivity of the ear are obvious.

We cannot put back the clock. Equal temperament provides the most effective compromise so far discovered. It is ignorance of the nature of this compromise which is inexcusable, as is the assumption that it should dictate to our ears where it has no right (*e.g.* in string playing or singing). This is like saying we should all conform to the measurements of ready made suits.

These comments may seem strange coming as they do from a professional pianist. Yet the effect on my understanding of the *quality* of intervals and chords has undergone a fundamental change since Mr Boyle tactfully left one of his unique calibrated monochords for me to experiment on. I do not think it possible to overestimate the importance of this device, either musically or educationally. It may be that we live in the best of all possible musical worlds, but this is for each one of us to decide in the light of our experience and such experience is not available without easy comparisons of the kind made from the examples given by Mr Boyle. I would most earnestly recommend anyone connected with music to make one and form their own conclusions. After all, dissonance and consonance are attributes of human hearing, not qualities inherent in sound waves in the air.

PREFACE

THOUGH primarily intended for students of music this book should also appeal to all those lovers of music having an enquiring turn of mind. Its treatment of the subject is a practical one based on the facts of musical experience and their influence on the history and development of music.

The name of Llewelyn Southworth Lloyd is sufficiently well known in the field of musical acoustics as to need no introduction from me. He is perhaps best remembered for his excellent articles in the fifth edition of Grove's *Dictionary of Music and Musicians* and for his text-book *Music and Sound*. In addition to these he wrote altogether, during and just after World War II, about forty-five articles on musical acoustics and intonation for various musical and scientific journals and periodicals, copies of which were then so limited in number by shortage of paper that they are practically unobtainable today. Hence the reason for their reappearance here in book form. In choosing from this total the articles contained in Part I of this book I selected those which I thought would have the widest appeal and which could be arranged, with the minimum of alteration, to form a reasonably logical sequence and coherent whole. Nevertheless, it should be borne in mind that these articles, written separately at different times and presumably intended to be more or less complete in themselves, must inevitably overlap in their treatment of some aspects of the subject matter. This however, may not be altogether such a bad thing in what is for most people a difficult subject. Finally, to round off the first part, I have added an introduction and bibliography, and have quoted throughout, in square brackets, the numbered items of this bibliography as reference to them occurs in the text.

The second part of the book has been written mainly to supplement the first.

Apart from the obvious, the 'Glossary of Terms used in Musical Acoustics' has been compiled with the intention of forging a common link between the separate articles and to provide a basis for further study. It is also an attempt to consolidate the work done by Lloyd on musical acoustics, since it is from his writings that the majority of the definitions in it have been derived.

It is hoped that all serious readers will make (or have made for them) the monochord illustrated in Fig. E and will use it to verify for themselves the basic facts of musical acoustics. In achieving this they may well be inspired to make further experiments and to discover and develop their individual taste in melodic and harmonic intonation.

Although dealing in the main with the purely theoretical aspects of the subject, Chapters 19 and 20 are of importance to those who wish to understand how and why the physical measurement, addition, subtraction, etc., of intervals can be carried out mathematically; what the aims and limitations of the various temperaments and tunings are; and so on.

The majority of people have little use for fractions and ratios in their everyday lives. The 'Summary of Mathematical Terms and Operations' is an attempt to jog their memories on a few important points concerning these things, and in addition, to introduce the concepts of logarithmic units and scales—both exceedingly useful tools in the study of musical acoustics.

Besides saving a lot of tedious calculation, the specially compiled table at the end of the book serves as a bridge between theoretical and practical musical acoustics. By means of it any frequency ratio, whether stated simply as an arithmetic ratio or in logarithmic units, can be translated directly into its equivalent string-length (ratio) on the monochord, hence to an aural experience—and vice-versa.

For permission to reproduce the articles contained in Part One I have first to thank the author's daughter Mrs A. M. Rowe. These articles have already appeared in much the same form under the same headings in the following musical journals: the first four in *The Musical Times*, the fifth in *The*

Music Review, the sixth and seventh in the *Monthly Musical Record*, and the eighth, ninth and tenth in *Music and Letters*. The editors of these journals are cordially thanked for giving their permission to reproduce these articles, full particulars of which are given in the bibliography on pages 112 to 114.

I wish also to express my thanks to all those library authorities and librarians through whose co-operation I was able to see original copies of most of Lloyd's articles. Many of these, I am told by my local librarian, were obtained through the South East Regional Bureau and the National Central Library.

I am indebted to Messrs J. M. Dent and Sons, and W. W. Norton & Co. Inc. for permission to reproduce Fig. 1 of Chapter 5 together with a number of quotations from Alec Harman's modernised edition of Thomas Morley's *A Plain and Easy Guide to Practical Music.* This edition was not available at the time that Lloyd wrote his article. His quotations were taken from the Oxford University Press facsimile edition of this book published for the Shakespeare Association.

Eric Blom's bibliographical note on Ll. S. Lloyd from the fifth edition of Grove's *Dictionary of Music and Musicians* is reproduced by kind permission of the publishers Messrs Macmillan & Co. Ltd.

As Lloyd points out (see Bibliography items 27 and 28), Ellis's translation of *Tonempfindugen* is not as accurate as it might be and I have to thank Dr. J. A. Boyle for providing me with a more accurate translation of some of its sections. It is now exactly a hundred years since this remarkable book first appeared.

I am indebted to William Coates for reading my original manuscript and for his helpful discussions with me regarding the text. My thanks are also due to Prof. Thurston Dart for his interest, support and advice. For the sketches of the monochord I am indebted to my colleague Frank Jarrett.

I am especially grateful to Kenneth van Barthold for providing a Foreword and for trying out the experiments on the monochord from their descriptions in the text.

Finally, I must thank the publishers and all those friends and colleagues who have helped in any way in the preparation and checking of this work.

H.B.

NOTE TO THE SECOND EDITION

The new material represents an increase of about forty per cent.

Except for some footnotes the first part of this book remains unaltered.

Much new matter has been added to the second part—particularly to Chapters 12, 13 and 21. Seven of the Appendices are entirely new.

A 'Further Reading List' has been added on p. 298 and its 'FR' numbered items referred to in the text.

All new acknowledgements have been inserted as the need for them occurred throughout the work.

One or two errors have been corrected—especially those in the mean-tone frequencies given in Chapter 20.

I had hoped to include some long-prepared Ratio Tables of greater precision but it was considered uneconomic to change the existing ones in this edition.

I wish to thank George Sargent most cordially for the interest he has shown in this work and for the encouragement he has given me over the years.

I am very grateful to Len Plumstead for reading through my new manuscript and for his assistance in the tedious job of proof-reading.

Lastly, I must once more thank the publishers for the opportunity they have given me of presenting a second edition and for their help and cooperation in producing it.

H.B.

CONTENTS

CONTENTS

PART II

by Hugh Boyle

Musical Acoustics

Including Definitions, Experiments, Measurements, Calculations, Various Appendices and Tables

PART I

ARTICLES BY LL. S. LLOYD

WITH AN

INTRODUCTION BY

HUGH BOYLE

INTRODUCTION

A BRIEF HISTORY OF THE SCIENCE OF VIBRATING STRINGS IN THEIR RELATION TO MUSIC

Pythagoras (570–504 B.C.) is usually credited with the discovery that a vibrating string, stopped at two-thirds or one-half of its length, sounds the fifth or the octave of the note it produces when vibrating freely; *i.e.* that lengths of ratio

$$1, \frac{2}{3}, \frac{1}{2}$$

would sound a note, its fifth, and its octave. In this way he and his followers were able to link up what they heard—the pleasing sensations perceived through their ears—with ratios of vibrating string-lengths—physical facts and occurrences perceived through their visual faculty and sense of touch.

The set of numbers quoted above bear a certain relationship one to the other which can be made more obvious by dividing them each by two. This in no way affects their ratios one to another and the series now becomes

$$\frac{1}{2}, \frac{1}{3}, \frac{1}{4}$$

The Greeks, who were devoted mathematicians, fully recognised the simple and orderly mathematical relationship connecting together any such a set of numbers and described them as being in harmonic proportion. It is important however to realise that this was a purely mathematical term having no connection whatever with the word harmony in the sense that it is used in music today.

The Pythagoreans ignored the fact that their ears alone had selected these particular intervals initially, and in accordance with their philosophy, which was dominated by mathematical conceptions, declared music to be a natural perfection, dependent upon numerical proportions, but quite independent of the ear. Consequently they did not consider it necessary to explain how it was that their ears were able to perceive or to compare vibrating string-lengths, or why they should recognise and prefer only those lengths whose ratios were formed from small whole numbers.

During the next two thousand years no important additions were made to our knowledge of the science of vibrating strings. Then came Galileo (1564–1642) and Mersenne (1588–1648). These two, working independently, discovered that the time taken for one complete vibration or cycle to occur—called the period of the vibration—was determined by the string's length, tension, and density. Thus, time—a new factor—was introduced, making a fresh approach to the problem possible. For whereas it had not been at all clear how the ear could estimate lengths, it seemed much more reasonable to suppose that the ear could detect differences between the sensations produced in it by air vibrations originating from strings whose complete vibrations occurred in different periods of time.

The next step was to find out the exact nature of the string's movement during any one of its cycles. This was not at all obvious, since the string's motion was much too fast to be followed by the eye. The discoveries of Wallis (1616–1703) and Sauveur (1653–1716) were therefore all the more remarkable. For each of them, working independently, discovered that a string vibrates, not only as a whole, but also and at the same time, effectively, as two halves, three thirds, four fourths—and so on. From the direct relationship already established between length and period of vibration, it then followed that every individual vibrating section would have a period of vibration proportional to its length, *i.e.*, that such a string would effectively produce a set of simultaneous sectional vibrations whose periods formed the ratio

$$1, \ \tfrac{1}{2}, \ \tfrac{1}{3}, \ \tfrac{1}{4}, \ \text{etc.}$$

—one complete vibration of the string, as two halves, taking only half the period of time of one complete vibration of the string, as a whole—and so on.

With this discovery came the explanation of another great mystery. Aristotle (384–322 B.C.) had recorded that the note of a vibrating string contained some element of its octave. Later experimenters heard more of these upper partial tones or overtones—as they are now called. Mersenne detected four, which he reported as occurring at the octave, the twelfth, the fifteenth, and the seventeenth of the fundamental note—though neither he nor his predecessors were able to offer any satisfactory physical reason for their existence. However, the discoveries of Wallis and Sauveur, mentioned above, brought new evidence from which a perfectly reasonable explanation was possible. For upper partial tones spaced at just these intervals, were exactly what would be expected to be heard from such a set of vibrations. Thus the note produced in the ear by a vibrating string was shown to be composed of a series of pure tones (partials), sounding simultaneously, and corresponding to component or partial vibrations of the main vibration whose periods (or string lengths) formed an harmonic series.

Nowadays, when referring to the physical properties of a vibrating string it is more usual to speak of the reciprocal of its period—termed its frequency. Since periods are measured in seconds per complete vibration, or cycle, then from its definition, frequency must be measured in cycles per second. It thus follows that if the ratios of the periods of a complex vibration form the harmonic series

$$1, \tfrac{1}{2}, \tfrac{1}{3}, \tfrac{1}{4}, \text{ etc.,}$$

then the corresponding frequencies must be in the ratio

$$1 \quad 2 \quad 3 \quad 4 \quad 5 \quad \text{etc.,}$$

—an arithmetic series.

This linking up of the vibrations of a string with a mathematical series was the starting point of much speculation by musical theorists that followed, the most important of these being the attempt to make it a basis for the 'explanation' of classical tonality and harmony. Both speculations left out the part played by the musical ear and completely ignored the history and development of the art of music.

Now although this leaving out of the ear would, at first, seem absurd—rather like ignoring the eye in the visual arts and allowing only those shapes or forms whose contours could be rigidly defined by reasonably simple geometrical figures or algebraic equations—on second thoughts, it is perhaps not really so surprising, since there is a strong tendency in most people to take the gift of hearing for granted whatever the circumstances. For instance, in driving through, or in crossing a busy street safely, the ear plays a far more important part than it is given credit for; and which would people choose as a permanent loss—supposing that such a choice was forced upon them—the sound or vision of their television set?

Helmholtz (1821–1894) was one of the first to realise this omission of the ear. Following up the work of Ohm (1787–1854), he reasoned that as sound was a sensation perceived through the ear by the brain, it was therefore quite illogical to expect that it could be wholly accounted for in terms of physics or mathematics, and that in fact what we hear must be very largely influenced by what our ears and hearing faculty can detect and perceive from the vibrations causing the sensation of sound; *i.e.*, that the physiology and psychology of our ears and hearing faculty must also be taken into account.

After testing his theories by means of a number of well chosen and carefully thought out experiments he was led to conclude that there was present in the mechanism of the ear some property which made it peculiarly sensitive to the beating which occurred between some of the several upper partial tones of two notes when sounded simultaneously, and that these beats generally produced an unpleasant sensation of roughness or dissonance in the ear having a maximum irritating effect when occurring at about thirty per second in the treble stave. Armed with this information he was able to show that the more consonant an interval was, the freer it was from this roughness, the tighter was it hedged in on both sides by dissonance caused by beating, and consequently, the greater was its 'definition' in the ear. This important discovery not only answered the question 'why does the ear prefer those intervals whose corresponding frequency or string-length ratios consist of small whole numbers?', but also showed that the order of consonance was determined (physically) by their degree of smallness, *e.g.*, the fifth, ratio $\frac{3}{2}$, is more consonant than, say, the major third, ratio $\frac{5}{4}$. In this way he gave to consonance a physiological meaning, separating it from concord, a purely musical term.

Helmholtz was also the first to show how the quality of any musical note depended only on the number, selection, and relative loudness of the partial tones which the mechanism of the ear could extract from the vibration causing it, and how difference tones could be generated in the ear by two or more strong vibrations occurring simultaneously. (See *quality*, Ch. 14.)

Since the time of Helmholtz work has been done on the relation between the sensations of sound and their corresponding physical causes,[1] particularly by Harvey Fletcher.[2] He demonstrated beyond any shadow of doubt that to assume the pitch and loudness of a note can be directly correlated with the frequency and intensity of the vibration causing it, as many have done before him, was quite wrong. He found that a change in the intensity of a sound vibration produced a change in the pitch perceived from it and also that a change in the harmonic content of such a vibration produced a change in both pitch and loudness perceived from it. The first of these effects can easily be verified using the B.B.C's tuning note—a pure tone of frequency 440 cycles per second—which is transmitted regularly by all B.B.C. transmitters before starting their normal transmissions. Adjust a receiver to give either a very loud or a very soft note, then, having 'noted' its pitch, re-adjust the receiver quickly to give the opposite effect. Under these circumstances the pitch of the note will be found to rise or fall by at least a semitone—in spite of the fact that the frequency of the vibration producing it will have remained perfectly steady throughout at 440 cycles per second. At 220 cycles it can be as much as a minor third.

In view of the above it is not surprising to find that when considering the note produced in the ear by a string which is allowed to vibrate freely, such as a piano string, things become very complex. For in this kind of vibration, known as a damped or natural vibration, not only does each successive vibration lessen in intensity, making the vibration as a whole non-periodic, but also the ratios of the relative intensities of its component vibrations alter continuously throughout its

[1] Fletcher & Munsen, *J.A.S.A.*, 5, 82, 1933.

[2] Fletcher, *J.A.S.A.*, 6, 59, 1934.

duration. A further complication arises due to the fact that the periods of these component vibrations do not fall exactly into the harmonic series and consequently they produce correspondingly inharmonic upper partial tones in the ear. For instance, R. W. Young[3] recorded that the frequency of the fifteenth sectional or partial vibration of one of the lower strings of an upright piano was almost a comma (22 cents) sharper than that corresponding with the 16th term of the harmonic series based on the period of its fundamental vibration. In such a case the pitch of the note produced does not give the same degree of 'definition' in the ear as does, for instance, the note of an oboe or of a bowed string. The note of the latter originates from a periodic vibration rich in harmonic partial vibrations, hence its well defined pitch, whereas, not only is the vibration of a piano string non-periodic, but its higher partial vibrations are purposely weakened. Finally, when considering the intervals obtainable from such an instrument, since the higher partial vibrations are weak and the vibration is damped, no appreciable difference tones can be generated in the ear, except possibly for a brief period between lower partials at the commencement of the notes forming the interval. Thus beating involving difference tones is practically eliminated and can therefore add little to the interval's 'definition' in the ear.

There can be no doubt that the slight uncertainty in the 'definition' of both note and interval in the ear from the notes of a piano helps to cover up the deficiencies of equal temperament and allows the ear sufficient freedom for it to take advantage of this uncertainty and perceive some degree of enharmonic change, if, for instance, it is suggested by the supporting harmony. In this connection Ll. S. Lloyd writes,[4] 'The notes of the piano rapidly diminish in intensity, more rapidly indeed than they do in loudness, which is a perceptual effect. As the note ceases to sound, things may well happen to our perceptions which our ears would not permit when the note was first struck'. This would seem to account, at least partially, for the fact that a change in the intonation perceived from the same notes—played simultaneously on the piano—

[3] Grove V., Vol. VII, p. 597—see also p. 145.
[4] [Ref. 11], p. 77.

can often by achieved by a carefully judged change in the relative weight allocated to the individual notes, and by almost imperceptible differences in the time at which each note is struck.

The exact process by which the ear analyses the effects on it of a damped vibration is still a mystery, since, as Helmholtz himself pointed out, his method can only apply to periodic vibrations. What is certain, however, is that our aural perception is given more latitude when dealing with damped vibrations than it is given when dealing with periodic ones.

General References

Helmholtz. *Sensations of Tone* 1875, Eng. Trans.
Lloyd, Ll. S. [Ref. 1, 11, 17, 24]—see p. 112.

I

THE PERFECT FIFTH

THE perfect fifth is at once the perfect concord and the perfect enigma of musical theory. If we start from a given note, and tune two series of intervals from it with theoretical accuracy, one in ascending fifths and one in ascending octaves, we shall never return to a unison. Yet the fifth, or the fourth as an interval approached downwards, is perhaps the one interval, other than the octave, which we may count on finding in widely different musical scales, evolved by different peoples, in different countries, and in different times.[1] How can this be possible if octaves and fifths cannot be fitted together? Now let us turn to the piano and, beginning at bottom A, play upwards from it, first twelve successive fifths, and then seven successive octaves. We shall arrive each time at the same note, the top A of the keyboard. Twelve fifths have now been fitted into seven octaves. Yet the Greeks discovered, nearly two thousand five hundred years ago, that twelve perfect fifths would exceed seven octaves by a small interval known as the comma of Pythagoras. Obviously, therefore, each fifth on the piano has been flattened by one twelfth of this interval. In our progress up the piano in a succession of fifths we touched on every note of the chromatic scale—once. Here, then, is the explanation of equal temperament, which makes F sharp the same as G flat, and so on. So simple and obvious is it that many theorists have convinced themselves (and some musicians) that, ever since the time of Bach (1685–1750), composers have thought in equal tempera-

[1] *E.g.* The Japanese scale described in the *Musical Times* for July 1939, p. 512.

ment. Some of them find that a soul-satisfying explanation. Others are much troubled by the conclusion that, since the middle of the eighteenth century, European music has rested on a false foundation. They comfort themselves by the thought that, after all, Beethoven (1770–1827) was a great composer.

The notion that composers and artists think in equal temperament leads to obvious difficulties, in music and logic. The violin is tuned in a succession of perfect fifths; and, taking advantage of the free intonation of their instruments, string-players, if they are skilled and possess the delicacy of ear they need for success, will persist in playing fifths which the trained musical ear accepts as perfectly in tune. Even the tolerant ear of the pianist can recognise the beauty of the intonation of a really good string-quartet. Then again, equal temperament is a comparatively modern invention. For four or five centuries before its use became generally established, a different temperament, called mean-tone tuning, was in vogue. Obviously, therefore, before Bach's time composers must have thought in mean-tone temperament! But what about still earlier composers? What temperament did Dunstable (?–1453) Fayrfax, Aston and Taverner think in, before temperaments had been invented? At this point some student of counterpoint, greatly daring, will suggest: Perhaps they didn't think in temperaments at all: perhaps they thought in terms of musical intervals: perhaps their successors did the same if they wrote contrapuntally: perhaps Parry (1848–1918) was right when he insisted that scales are made in the process of endeavouring to make music[2]: perhaps Stanford (1852–1924) was right in distilling the pure scale, with some of its notes moving about, from the counterpoint of Palestrina (1525–1594) and our Tudor composers.[3] And, encouraged by his success, the contrapuntist may even suggest that perhaps Bernard van Dieren (1884–1936) was right in thinking that 'the alleged problems of euphony that obsessed theorists solve themselves in well-balanced polyphony.'[4]

At this point the man of science will doubtless take up the

[2] *Art of Music*, p. 16.

[3] *Musical Composition*, Chapter 2.

[4] *Down among the Dead Men*, p. 226.

rôle of prosecuting counsel. He may ask: How can a string player or a singer produce equal temperament *exactly* when the piano-tuner, with much better opportunities, cannot do so? Do you know (he will continue) that scientific instruments are now available which can be used to record, rapidly and accurately, the rate of vibration of every note of the piano? If the tuner could tune the piano exactly in equal temperament, the graph of these different vibrations would be a straight line. Actually, it is a waggly one, showing that some notes are slightly flat and others slightly sharp (see p. 75). If once you admit that artists do not play *exactly* in equal temperament you have given your whole case away. You have opened the door which allows them to make enharmonic changes of intonation, and so to play in tune. Why do you confuse the issue, he will continue, by leaving the ear out of your picture? Why not consider the varying degree of accuracy with which the human ear estimates the different musical intervals? And is there no musical evidence of that enharmonic latitude in the modern twelve-note semitonal scale which is the explanation offered by Dr. A. F. Barnes,[5] an explanation which fits all the facts, scientifically. Surely your notion is as faulty, scientifically, as it is unsound musically.

These reflections are prompted by the first of the 'Answers to Correspondents' on p. 523 of the *Musical Times* for July 1939. 'Cunedros' ended his questions with one concerning the '"perfectness" of the perfect fifth': a pertinent question for any understanding of the scale system of European music, but one which requires for answer far more space than can be found in the columns of 'Answers to Correspondents'. The fifth comes striding into the problem of the scale with irresistible momentum. Here are two musical concords:

In two-part counterpoint, sung or played in perfect tune, the D in the first one will be a perfect fifth above the dominant

[5] *Practice in Modern Harmony*, p. 29.

of the key of C; the A in the second one will be a true major third above the subdominant. If this A and D are then sounded together they will produce a dissonance. They will be separated by a flattened fifth. Its degree of flatness will be very noticeable—being, theoretically, eleven times that of the fifth of equal temperament. One of these two notes must move if the interval is to sound perfectly in tune. That is why Stanford describes the second-degree note of the pure scale as a mutable note. Why has the perfect fifth such masterful importance in intonation that it makes the musical scale into a flexible thing? Is there any reason, in nature, why counterpoint should distinguish the fifth and the octave as perfect concords, and the thirds and sixths as imperfect ones?

It was Helmholtz who first provided a scientific answer to this question by calling attention to the part which the ear must play in musical theory. We shall find no answer in nature, if we limit 'nature' to the vibrations in the air outside our ears. The answer depends on the properties of our ears and the sensations produced in them by the sympathetic vibrations of their nervous appendages. For an understanding of Helmholtz's answer we must begin with the series of harmonic overtones to be found in the sensations which the vibrations of strings or organ pipes excite, in this way, in our ears. The first six of this series of harmonics (a fundamental and its overtones) for the note C in the bass clef may be represented in musical notation as shown in:

It has been known for about two hundred and fifty years, as the result of scientific experiment by Mersenne, Galileo, Sauveur, and others, that the second harmonic (middle C) counting from the bottom corresponds to a vibration which is twice as fast (*i.e.* has twice the frequency) as that corresponding to the fundamental (C in the bass clef); the third harmonic (G of the treble clef) to one vibrating three times as fast; the fourth to one vibrating four times as fast; and so on. Also we may think of the harmonics as becoming fainter and fainter as we ascend.

Helmholtz's explanation of dissonance between two notes sounded together was that it is due to beating between their respective harmonics. This is shown in musical notation in:

wherein the notes forming the interval are represented by white notes, and their overtones by black ones. In each case the overtones are continued sufficiently far to reach a unison. The intervals chosen are the two perfect concords, the octave and the fifth, and two imperfect ones, the major third and sixth. Now suppose the intervals to be slightly mistuned. The unisons between the harmonics will also be mistuned, and they will beat; and since the lower harmonics are normally the most powerful ones, the beating will be more noticeable with the octave and the fifth than with the two imperfect concords. This makes us conscious, at once, of mistuning of octaves and fifths. We are less conscious of mistuning of the thirds and sixths because the disturbed unison is between fainter harmonics. For that reason the octave and fifth are said to be sharply defined.

Observe, too, that in the harmonics of the imperfect concords there is, in each case, one pair a tone or, worse still, a semitone apart. For the major third they are C and B, for the major sixth G and A. These will produce beats which will tend to hide the beating between the fainter harmonics that have ceased to be unisons. That is a further reason why the imperfect concords have less definition than the fifth or the octave. As intervals they have vaguer outlines, while the outlines of discords are vaguer still. This is perhaps the explanation of the '"perfectness" of the perfect fifth' which the correspondent sought, and it will doubtless interest others.

2

THE MAJOR THIRD

THE fifth is perhaps the most important interval in musical theory; but, in some ways, the major third is the more interesting one. Four major thirds may be catalogued. The first is the consonance, giving the accurately tuned interval which is, ideally, the concord of strict counterpoint. The second is the major third of equal temperament. The third is the Pythagorean interval produced by the sum of two major tones: the theoretical interval of the Greek (melodic) modes corresponding to our major third. The fourth is the indeterminate interval we produce in counterpoint when we treat a third, between two of the parts, as an interval that produces an 'unessential note,' *i.e.*, a note which is not part of the prevailing harmony, as shown by the bracket at Ex. 1:

Ex. 1

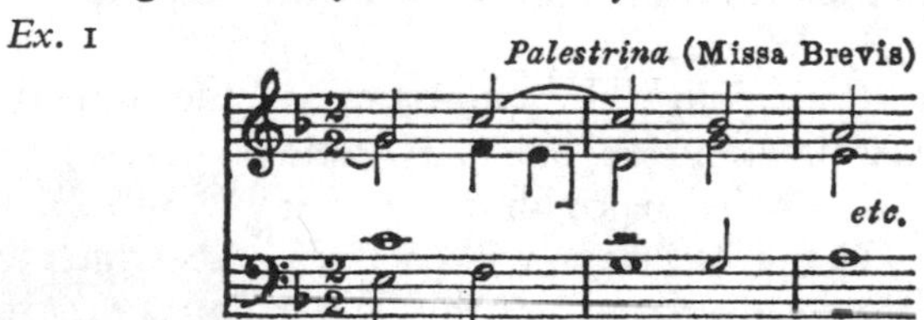

In all these descriptions we are thinking of the physical relations of the vibrations in the air which correspond to the tones and intervals we hear. Modern science tells us that we cannot assume exact correspondence between these vibrations and the tones we hear, save in the case of a pure consonance defined by the absence of beating between harmonics which was described in Chapter 1. We might therefore add to our catalogue 'The Thirds we Hear'; but that is another story which is told briefly in chapter 9. In the present chapter, lack of correspondence between the vibrations in the air and the tones we hear will be considered only to the extent that is indicated by the lack of aural definition of the interval, the attribute which leads us to describe the octave and fifth as 'sharply defined'.

Having reached this point the musical reader reckons that this is going to be a terribly stiff article to read. He may be assured that the worst is already over, and that the moral of the reference to modern science in the preceding paragraph is just this—that a knowledge of sixteenth-century counterpoint is more important in musical theory than a knowledge of physical acoustics; for it is the touchstone by which to test the conceptions of theorists who think primarily about the vibrations in the air instead of about the music we hear.

Let us examine our four thirds in turn. The first one, the perfect consonance, is the interval produced between two notes when the fifth harmonic of one is a unison with the fourth harmonic of the other. It is represented in Ex. 2 which appeared in the previous chapter. The white notes for each interval are supposed to be sounded together. Since the fifth harmonic corresponds to a vibration whose rate is five times

Ex. 2

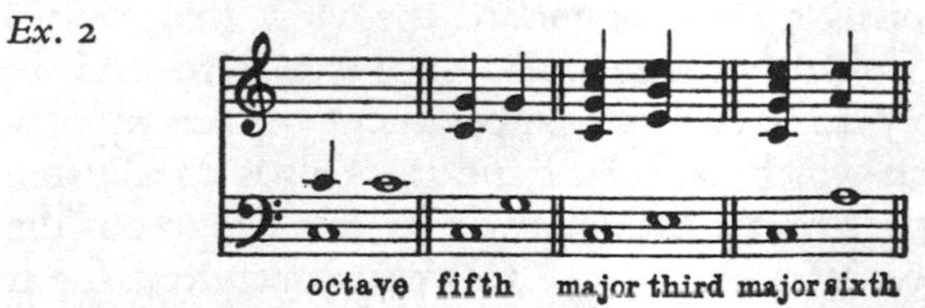

that of the fundamental vibration, and the fourth harmonic to one whose rate is four times that of the fundamental vibration, it only requires simple arithmetic to discover why, in a perfectly tuned third, the rates of vibration corresponding to the two notes which form the interval (here C and E) must be in the ratio $\frac{5}{4}$.

The major third of equal temperament is a larger interval than the true major third. It is the considerable mistuning of the thirds and sixths which is the chief defect of this method of tuning a keyboard instrument. In the previous chapter the construction of the chromatic scale of equal temperament was illustrated by a series of twelve fifths reached in succession from the bottom A of the piano. The reader will recall the wearisome repetition of fifths and fourths by the piano-tuner. Since the fourth is the inversion of the fifth, what the tuner does, in effect, is to fit twelve flattened fifths into seven octaves. But what the musician wants to know is: What kind of a 'scale' does this produce? The best answer, perhaps, is to draw a picture of tuning in equal temperament side by side with pictures of true diatonic intervals. Such a picture is given in Ex. 3 opposite.[1] In each half of this figure the intervals are drawn with theoretical accuracy, ignoring any vagueness in these same intervals, as heard, which might be attributed to their lack of 'definition'. It will at once be observed that the whole tones of equal temperament are all equal, and exactly twice as big as the tempered semitone. The sharpness of the major third is very evident. On the other hand, the minor third, between E and G, is too flat. Quite a substantial interval has been cut off the bottom of it, and the flattening of the fifth cuts a correspondingly small piece off the top of it. An error of precisely the same amount is added to the major sixth from C to A, as is of course inevitable since a minor third and a major sixth make up an octave.

Some theorists have propounded the view that, since the perfect fifth is sharply defined, and since true thirds and sixths are less sharply defined, equal temperament tampers most with those intervals which can best stand it. This is an illustration of the old adage that a little knowledge is a dangerous thing. The definition in question, due to unisons between the harmonics (Ex. 2), is only concerned with the interval between *two* notes. The theorist's conclusion, quoted above, applies only to *two*-part counterpoint. The musician will at once put him right. He will say: 'What I particularly dislike about equal temperament on the harmonium [there are good acoustical reasons why it is better suited for tuning the piano] is the false intonation of the thirds and sixths'. And if he happens to have heard a harmonium tuned in mean-tone temperament he might continue: 'So long as we stick to

[1] The writer is indebted to the Oxford University Press for permission to use this figure, and similar figures throughout the book, which have been adapted from his *Musical Slide-Rule*. [Ref. 3.]

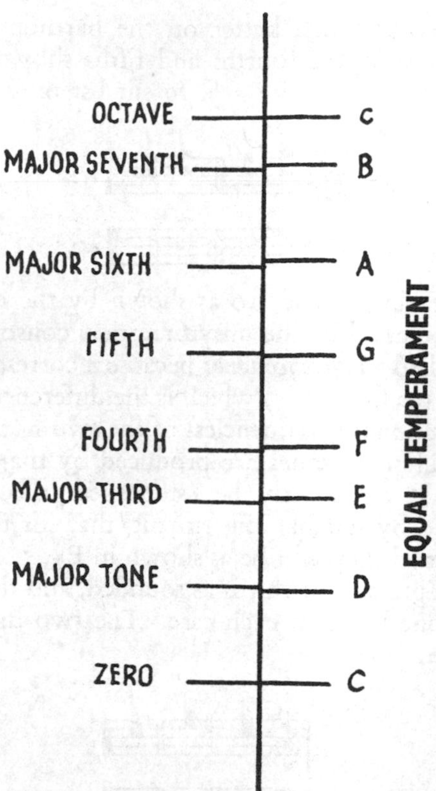

the keys with not more than, say, two sharps or flats, the major thirds of mean-tone temperament are perfectly in tune, and the minor thirds far better than in equal temperament. Surely, if equal temperament had the merits you claim for it, it would have displaced mean-tone tuning long before it did; in fact, it would have prevented its adoption. How do you account for the fact that it didn't?' And he would be perfectly right.

Helmholtz supplied the knowledge which our theorist lacked, or failed to apply, and it is interesting because it depends on the properties of our ears. If two notes less than an octave apart are sounded together, *loudly*, a deeper note, much fainter, is also heard. It is not easy to detect this at the piano because the tones of any percussion instrument fade rapidly; but here is an experiment which works on most pianos. It works much better on the harmonium. Play in succession, *andante,* the fourths and fifths shown as crotchets in Ex. 4, and strike the keys *ff*. Intent listening will reveal a

Ex. 4

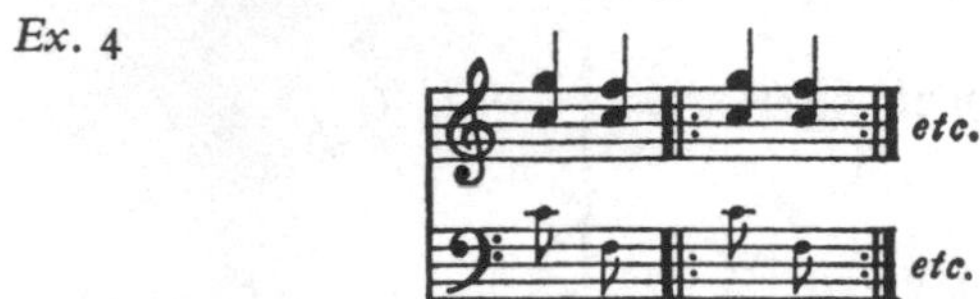

faint tone leaping to and fro as shown by the quavers. This faint tone is created by the unsymmetrical construction of our ears. It is called a *difference tone;* because it corresponds to a rate of vibration (or frequency) which is the difference between the rates of vibration (or frequencies) of the two notes sounded.

Similar difference tones are produced by major and minor thirds; and it can readily be established, theoretically, and demonstrated by suitable experiment, that for the true thirds in a major triad they will be as shown in Ex. 5. The crotchets in 1 and 2 represent the thirds as sounded, and the quavers the difference tone heard in each case. The two difference tones are the same:

Ex. 5

and they will coalesce if both thirds, perfectly in tune, are sounded together as a major triad (3). Now if an interval is made sharp (*i.e.*, larger) the difference between the rates of the vibration of the two notes increases. The pitch of the difference tone will therefore rise. For example, suppose the rates of vibration of two notes were 250 and 200. The rate of vibration corresponding to their difference tone would be 50. But if the first note were sharpened, say, to 255, the rate of vibration corresponding to the difference tone would become 55. The pitch of a note corresponding to 55 is obviously higher than the pitch of one corresponding to 50 (vibrations a second). Similarly, if an interval is flattened (*i.e.*, made smaller), the pitch of the difference tone will fall.

Now think of the major triad, and suppose the middle note to be sharpened. This makes the major third bigger and the minor third smaller. One difference tone (the first quaver) will rise in pitch, and the other will fall in pitch. They will beat! And it happens that, for each of the two thirds, the harmonics which ought to be unisons are also beating, the one pair about twice, the other pair about three times as fast as the two difference tones. Quite a volume of dissonance, due to beating, is therefore produced by mistuning this one note of a major triad. That is why in harmonic use, or in counterpoint in three or more parts, the third of a major triad in equal temperament offends the sensitive ear.

The minor triad is interesting. Transpose the major third in Ex. 5 up a minor third, and the minor third down a major third. We obtain Ex. 6:

Ex. 6

The difference tones are a fifth apart, and no mistuning of this fifth can make them beat, since they are pure tones. This helps to explain why the minor triad of equal temperament is not so harsh as the major one. But whether the intervals are true or tempered, the difference tones are alien to the harmony of the chord, and help to give it the 'veiled' effect we associate with a minor triad.

The remaining thirds we must dismiss briefly. The Pythagorean third is much too sharp to form a consonant triad on the tonic—about half as sharp again as the major third of equal temperament. But the modes of which it was characteristic (as in Greek theory) were melodic modes; and between notes sounded in succession there can be no beats. Pythagorean tuning is produced on the open strings of the violin. As was shown in the previous chapter, the interval between A, as the major third of the subdominant, and D, as the fifth of the dominant, is not a true fifth. But this interval is tuned true on the violin. If we add the C string of the viola we have the

Pythagorean intonation for G, D, A and E, with C as tonic. Further fifths at each end give us F and B. This tuning, with its octaves, makes the semitones very narrow. Curiously, it appears that string players, who instinctively play true intervals for the concords of two-part music, tend to produce Pythagorean intervals in unaccompanied melody.[2] Yet this is in no way surprising when one recalls the influence which the perfect fifth exercises even on melodic scales, and the tendency of the violin player to 'close' the semitone in a melodic passage.

What of 'unessential' thirds? Here the answer is simple, for the scientist can tell us little. The musical ear looks forward to the essential note to which the melodic line is moving. On a stringed instrument, the intonation of passing notes, reached on the way to the goal, is a matter of musical taste and a sensitive ear tempered by skill in performance; and that is about all that can be said of it.

[2] This contrast of intonations, in the performance of the same players, was established seventy years ago by Cornu and Mercadier (Comptes Rendus, Feb. 1869, quoted by Helmholtz). Recent measurements, made in America by P. C. Greene, confirm their results for unaccompanied playing, to which his investigation was limited.

3

THE PERCEPTION OF SMALL INTERVALS AND BEATS

There is an interesting experiment which organists frequently perform, though the significance of the result they observe does not become fully evident till it is related to other observations made in the laboratory. They would explain to the rest of us that the *voix celeste* is a stop consisting of a series of pipes each of which is tuned a trifle sharp on the corresponding pipes of an associated stop. When the two stops are used together they produce a sensation of beating which is not unpleasant, within limits, as it gives a wavy effect to the sound, suggesting, perhaps, the vibrato of a violin. The interesting thing about this combination of stops is the ear's response to it.

The first thing we observe is that the ear does not distinguish two separate notes when we press down a key, although vibrations of slightly different rates (frequencies) are being produced in the air. The ear hears only a single note. Thus we have obtained evidence that there are limits to the ear's power of separating two notes which have nearly identical rates of vibration. If, step by step, the interval between the two notes were made larger, we should each of us reach a stage at which we became conscious that we were listening to *two* notes very close together. The stage at which that would happen would not be the same for the ears of each of us. Until that happens the only evidence of a mistuned unison of which we are conscious is the beating described above. Such beating becomes unpleasant as, with increasing mistuning, it becomes more rapid. By contrast we may think of the effect of pressing down two keys a semitone apart using only one stop. We not only hear a most dissonant sound: we actually hear two notes.

The second observation is this. If, when we are using the two stops in combination, we push in the *voix celeste* stop, we become conscious of a very slight fall of pitch. Some ears will be more conscious of this than others. Some indeed may not be able to perceive it with certainty. But for those whose ears are more sensitive it appears, therefore, that the note attributed to the combination of stops is not the note of one of them heard singly. That this actually happens is confirmed by laboratory experiments. When two different notes are sounded together which the ear cannot separate, the pitch of the note heard lies between the actual pitches proper to each individual note.

There is a third observation. We can detect a fall of pitch when the *voix celeste* is pushed in. Our ears can therefore distinguish between the pitches heard, first before it is pushed in, and then after it has been pushed in. So we are not surprised to be told of laboratory experiments[1] which show that the ear can detect an interval between notes sounded in succession even though they may be too close together for the ear to separate them when heard together.

The beating we hear when the *voix celeste* is used with its companion stop depends on the rate of the physical beat in the air. For each key we press down, this rate is equal to the difference in the rates of the two vibrations which are produced in the air. Let us suppose that in the neighbourhood of middle C there were two vibrations whose rates were 260 and 264 cycles per second, respectively. We should hear 4 beats a second, but only a single note. Now suppose we took two notes an octave higher. Their vibrations would be twice as fast; in other words, at rates of 520 and 528 cycles per second. We should hear 8 beats a second, though the physical interval (which we cannot perceive) between the two beating notes would be the same at each octave. Reduced to its lowest terms, the ratio of this interval would be $\frac{65}{66}$, which shows that it is rather larger than a comma.

On the other hand, if the rate of beating were the same at the octave, *i.e.* 4 beats per second, the rates of the vibrations would only differ by 4. Let us say they would be 520 and 524.

[1] *E.g. Accoustics*, Alex. Wood, 1940, p. 490.

In other words, the physical interval between the beating tones would be, as nearly as may be, half what it was an octave lower. It may interest some reader to estimate the rate of beating with the *voix celeste* at different octaves, and so determine how rapidly the physical interval between the note of the *voix celeste* and that of its companion stop diminishes, on his organ, as he ascends the keyboard (remembering always that we cannot *perceive* this, as an interval, because our ears blend the two notes).[2]

The ear's power of detecting a difference of pitch between two notes *sounded in succession* has been very fully examined by research workers in certain American laboratories.[3] Their results disclose considerable differences between the ears of different people, as we should expect. Some people are able to detect small differences of pitch of which others are quite unaware. The ear is most sensitive for intervals between high notes, say those above the treble stave. As we proceed, successively, to lower and lower notes, the interval has to be gradually increased before the ear is conscious of a difference of pitch. We are reminded that, for faint sounds, the ear is most sensitive at a high pitch. Thus a pure tone, that is a tone free from overtones, with a (physical) intensity of, say, 30 decibels above zero, whose pitch lies in the two octaves above the treble stave, is readily perceived by the normal ear. But a pure tone of the same intensity in the deep bass is quite inaudible.

It is the sensation of beating that alone enables the ear to detect the existence of a small difference of pitch between two notes which it cannot separate *when they are sounded together*. Moreover, this sensation comes into play with two notes sounded together between which the normal ear could discover no difference of pitch even if they were to be sounded in succession. But why do we hear beating? Here the scientist turns to its cause, a physical beat in the air. When two vibrations in the air have slightly different rates they are constantly falling out of step, as it were, and then getting into step again.

[2] At the bottom of the treble stave, 4 beats per second would represent an interval of a comma.

[3] The most complete work is that of Shower and Biddulph. *J. Acoust. Soc. Amer.* (1931), 3, 275.

The result is that the physical intensity of the combined vibration in the air is constantly decreasing and then increasing. This ebb and flow in the intensity of the vibration is the physical beat.

The sensation of beating in the ear must be distinguished from the physical beat in the air. If the beat is slow, say, not more than 4 beats a second, the beating in the ear is not unpleasant. This happens with the *voix celeste*. But when the beat becomes rapid, say, 30 beats a second, the beating becomes very unpleasant. (This figure refers to notes in the treble stave.) If the beats become more and more rapid, the beating in the ear eventually ceases. The beats are then too rapid to be perceived. A similar thing happens in the cinema. We know that in fact there is a flicker on the screen; but, with modern apparatus, it is too rapid for the eye to be aware of it. The interval which produces 30 beats a second is about a semitone in the middle of the treble stave. But, as we should expect, the interval which produces the most unpleasant beating in the bass stave is more than a semitone.

To keep our minds clear we had better think now of pure tones, free from overtones. These tones correspond to perfectly simple vibrations, that is, steady to-and-from motion, in the air along a line pointing in the direction in which the sound is travelling. If the interval between such pure tones is more than about a minor third, in the treble stave, the beats are too rapid to hear and there is no beating in our ears. That the rate of the beats in the air is the difference between the ratio of the two vibrations which produce them is fairly easy to understand when the ratio between the rates of the two vibrations is given by successive numbers, as with the semitone, ratio $\frac{15}{16}$, or the minor tone, ratio $\frac{9}{10}$, or the major tone, ratio $\frac{8}{9}$. The vibrations then fall into cycles which repeat at the same rate as the beat. It is not quite so easy to understand when the rates of the vibrations have a ratio such as $\frac{9}{11}$, that there would be two beats in each cycle.

The scientist explains this at once by writing down a mathematical equation, but that means nothing to many of us. We can all understand his explanation, however, if he draws pictures of the vibrations for us. Unfortunately, the musician who is unfamiliar with such pictures is apt to fight shy of them.

But, in fact, if he will take a little trouble, any musician can understand them just as easily as he tells the time by a clock (an achievement which came only with practice and familiarity). Take a piece of cardboard and lay it on the table before you. Take a pencil in your right hand and hold the cardboard quite still with your left hand. Now lay the pencil point on the cardboard, and move the pencil away from you and towards you along a straight line pointing away from you. If you imagine that, say, one of your waistcoat buttons were able to sound a musical note, free from overtones, you are now drawing a very exaggerated, slow-motion picture of the vibration it would be producing in the air in front of you. Now stop and look at this picture, which will be a straight line. It tells you practically nothing about the vibration it is meant to represent, nor does it show where your pencil point had been at any given moment. To tell you all this, the scientist has a simple plan.

Hold the piece of cardboard in your left hand, and while continuing to move the pencil point steadily away from you and towards you along a straight line, pull the cardboard steadily from right to left. *Now* look at the cardboard and the curve you have drawn on it. That curve resembles a sound-curve, and it is just a handy diagram to show how you had moved the pencil. If you moved the pencil point steadily to and fro, like short swings of a pendulum, the curve on your cardboard will be a smooth curve. Any selected point on the curve tells us just where the pencil point was at some moment of the experiment. To find out at what moment this happened all we have to do is to measure how far the selected point lies to the right of where the curve began. *It is of little use merely to read about this.* You should take pencil and cardboard, and try the experiment. Sound-curves will cease to be mysteries (see graph p. 233).

Here the top curve is the simple smooth curve corresponding to a pure tone—sometimes called a sine curve. It is just like the curve you will draw on the cardboard with a little practice. If you will count the complete waggles in it, or the number of wave crests, you will find there are 9 of them. The same curve is repeated in the middle of the figure, but with it there is shown another curve, rather like it, with a dotted line. If

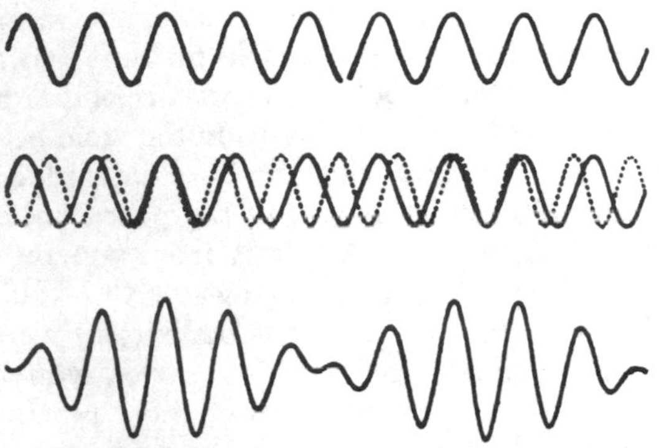

you count the complete waggles in the dotted line or the number of crests, you will find that there are 11 of them. We have drawn the sound-curves for two pure tones with vibrations in the ratio $\frac{9}{11}$. Nine vibrations of one fit into eleven vibrations of the other, and the two together show just one complete cycle. But there will be two beats in the cycle.

This is shown by the sound-curve at the bottom of the figure, which is a picture of the complex vibration in the air produced by combining the two vibrations in the middle of the figure. They are combined quite easily. If at any moment they would both push a particle of air (represented by our pencil point) in the same direction we add the displacements. If they would each push it in opposite directions we find the difference between the two displacements. Compare the result with the two separate vibrations, and you will be able to see how this happens.

Twice in each cycle the movements of a particle of air, pictured by the combined sound curve, lengthen; twice they become very short again. Each time they lengthen the energy, or intensity, of the vibration increases. Each time they become very short again the intensity of the vibration diminishes. This pulsation of the intensity is the physical beat. If the rates of vibration of two notes are in the ratio $\frac{9}{11}$, the musical interval between the notes is just a little larger than a minor third. But we are thinking of two notes in the neighbourhood of middle C; and at that pitch an interval of a minor third or more, between pure tones, produces sensible beating in our ears. This sound-curve therefore represents beats we can hear.

There is one thing to add to complete the story. These are pictures of pure tones, and—at any rate for notes in the upper half of the treble stave—we hear no beating between them if the interval is more than about a minor third. But on a musical instrument, such as an organ, we hear beats if an octave or a fifth is mistuned. Is there an inconsistency here? The answer is No! The notes of the organ are not pure tones. They contain overtones, and it is these overtones that produce the beating we hear when an octave or a fifth is mistuned. We may recall the musical example used in Chapter I.

The white notes in each interval are supposed to be sounded together but they are set down side by side so that the overtones of each can be shown above them. The overtones are continued for each interval till we reach a unison. If the octave, the fifth, or the major sixth is mistuned, the unison is mistuned and we hear beating, although the fundamentals themselves are so far apart in pitch that the physical beats they produce in the air are too rapid to produce beating in our ears. At the pitch in the bass shown for the major third, however, the fundamentals would themselves beat slowly enough to produce beating in our ears, whether mistuned or not; and this is why a third at so low a pitch sounds unpleasant and muddy.

4

THE ARTIFICIAL SCALE

'We must distinguish carefully between composers and theoreticians'.— Hermann von Helmholtz.

THE artificial nature of the scale constructed by nineteenth-century 'theoreticians' from arithmetical calculations which began *and ended* with the rates of vibrations (frequencies) is not always realised. Least of all was it realised by those 'theoreticians' who misled themselves by giving a pseudo-scientific colour to their calculations. They would then call their artificial scale the natural scale; though a musical scale with a rigidly fixed intonation would be most unnatural. Or they might call it just intonation, which would be self-contradictory since, if it is tested as a musical scale, its rigidity is found to make some of its intervals unjust, as we shall see. Some have even been known to call it the pure scale, which shows that they were unfamiliar with the pure scale of the polyphonic period, as described by Stanford. Fundamentally, their mistake was that they confused a musical scale with the tuning of a keyboard instrument. Calculation of the rates of vibrations is the proper approach to such tunings, for the basis of correctly tempered tuning is the estimation of beats, while the basis of a justly tuned fifth, as used to tune the violin, is the disappearance of beats. The 'theoretician' whose scale was firmly rooted in arithmetic was thinking of tunings, that is of temperaments, not of musical scales.

A discriminating account of the construction of what he called 'the artificial scale' was given by Donkin in his *Acoustics* (1870). His explanation was to the following effect.

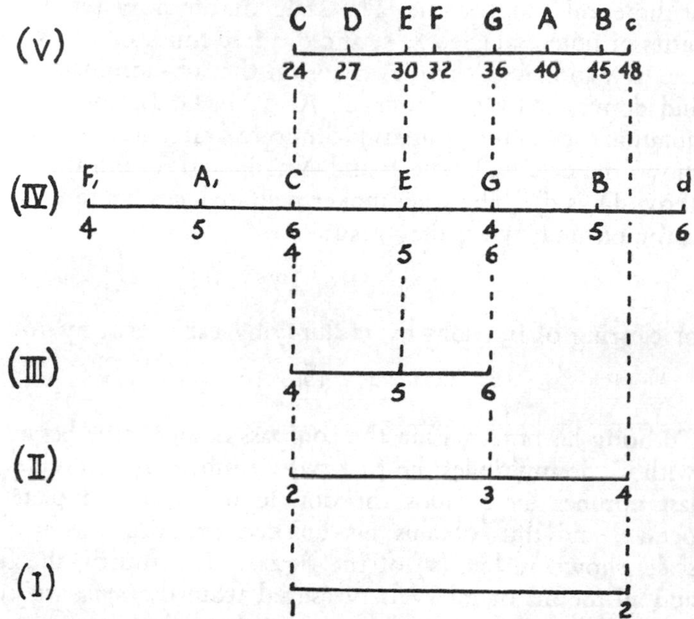

The scale maker begins with two notes an octave apart, shown graphically at the foot of the figure[1] as the two ends of a line marked 1 and 2 to give the ratio of their rates of vibration. He then moves them up an octave, when the ends of his line will be marked 2 and 4, as shown in line (ii). He observes: 'Between the two notes marked 2 and 4 there is a point of "natural bisection"[2] at 3. So we will mark this as the next most important note of the scale'. He now takes the longer section, marked 2 and 3 at either end, and moves it up an octave, when its ends will naturally be marked 4 and 6, as in line (iii). 'Between 4 and 6', he then observes, 'there must be a point of natural bi-section at 5. Let us mark it as another important note of the scale'. This device of natural bi-section is nothing abstruse; it is merely a new label for the familiar process of finding the arithmetical mean of, say, 4 and 6, which produces three terms, 4, 5, and 6, of what books on algebra call an

[1] The writer is indebted to the Council of the Royal Musical Association for permission to use this figure which appears in the *Proceedings*, Session LXVI, p. 35.

[2] This is the term used by Donkin.

arithmetical progression. The scale maker now sets out his series of notes, labelled 4, 5, and 6, three times in a row, as in line (iv), and labels them as triads on the sub-dominant, tonic, and dominant by the letters F, A, C, E, G, B, and D. (The notation used in the figure is Helmholtz's pitch notation, which shows the octaves below F and A as $F_{,}$ and $A_{,}$ and the octave above D as d). The scale maker next reduces his figures to a common unit, when they read:

$$1, \quad 5/4, \quad 3/2, \quad 15/8, \quad 9/4, \quad 45/16, \quad 27/8$$

or clearing of fractions by multiplying each term by 16:

$$16, \quad 20, \quad 24, \quad 30, \quad 36, \quad 45, \quad 54$$

To bring his notes within the compass of an octave beginning with C, he multiplies the first two numbers by 2, divides the last number by 2, adds the double of 24 to complete the octave, and thus obtains his finished product, the artificial scale, shown in line (v) of the figure. The fourth, the fifth, and all the major intervals measured from the *tonic* are justly intoned; but the artificial scale is a theoretical tuning, not a musical scale, for its rigidity makes some of its intervals, *e.g.* D to A,[3] unjust. It begins and ends with arithmetical calculations; it leaves the ear out of the picture, and as Donkin remarks:

> This proposition is to be understood merely as a statement of a mathematical fact, and not as involving any theory of the actual derivation of the scale.

Actually this is an understatement, for it would be just as logical to go through the same process with the shorter portion of line (ii) and so produce a perfectly ridiculous scale.

Many convinced 'theoreticians' of Donkin's day, reading his comment, would have protested that they were not engaged merely in arithmetical calculation. They would have explained that they took the numbers 4, 5, and 6 from the harmonic series. But that would have been to make matters worse by misleading any musician who did not know, as Thomas

[3] The ratio of this interval is $\frac{40}{27}$ which is less than $\frac{3}{2}$ the ratio of a perfect fifth such as C to G; see chapter 1, p. 12.

Morley (1557–?) knew,[4] that the harmonic series is only an arithmetical series. Consider the device of 'natural bisection'. This produces three terms, 4, 5, and 6, of what algebra books call an arithmetical progression. These terms correspond to *rates* of vibrations. If we invert them, we shall obtain terms corresponding to the *periods* of vibrations.[5] We shall then have three terms of what algebra books call a harmonical progression:

$$1/4,\quad 1/5,\quad 1/6$$

which conform to the explanation Morley gives of numbers in *harmonical proportion*. The mathematical definition of harmonical proportion, which Morley states, is most conveniently illustrated if we clear these terms of fractions by multiplying each by 60. They then become 15, 12 and 10. The difference between the first two is 3, that between the last two is 2. Morley would explain that these three terms are in harmonical proportion because the ratio between these differences, $\frac{3}{2}$, is the same as the ratio between the first and last terms, $\frac{15}{10}$. Morley understood, what the 'theoretician' did not realise, that when we take three successive numbers in harmonical proportion we are doing something which is the exact arithmetical counterpart of 'natural bisection'. The 'theoretician's' protest, with which this paragraph began, falls to the ground.

When this was pointed out to him, he might still have argued that the fourth, fifth, and sixth harmonics[6] of a musical tone form the common chord. But if we know anything of modern scientific work we can only tell him that he is mistaken. The pure tones which are the three harmonics of his arbitrary choice do not form a chord at all in the musician's sense. The musician's chord consists of 'notes' sounded by musical instruments. These notes are heard, not as pure tones, but as complex tones containing a series of harmonic overtones.

[4] See the Annotations to his *Plaine and Easie Introduction to Practicall Musicke* (1597), written before Mersenne had established (1636) the laws of vibration or strings, and long before Robartes announced (1692) his discovery that the notes of the trumpet are what we call harmonics, like those of a violin string.

[5] Morley's explanation was in terms of the lengths of strings, not of their periods of vibration, but the result is the same.

[6] See chapter 1, p. 12.

Such harmonic overtones give definition[7] to the 'notes', which enables the ear to hear them independently of each other. But the pure tones of the supposed 'chord of nature' are completely blended by the ear; their effect is to intensify its response to the fundamental of the tone to which as harmonics they are proper. They cause the tone, as a whole, to sound louder; they affect its quality; they may even affect its pitch if the tone is very loud; but they are not heard independently.

Finally, as a despairing effort, the 'theoretician' might have said: 'But experiment shows that vibrations in the ratio $\frac{5}{4}$ are heard together as a major third, and those with a ratio $\frac{3}{2}$ are heard together as a perfect fifth. Therefore if we hear a common chord it must be produced by notes with vibrations in the ratios 6 : 5 : 4'. He would then have completely defeated himself. First of all, modern Science would tell him, it is true that if two musical tones are produced by vibrations whose frequencies (*i.e.* rates of vibration) have the ratio $\frac{5}{4}$ they are heard together as a major third, at any rate if they contain enough harmonic overtones to give them definition. But the converse is not *necessarily* true. When we hear a major third in a musical performance we can only infer that the fundamentals of the tones we hear correspond to frequencies in the ratio $\frac{5}{4}$ *within the limits of accuracy with which the listening ear can estimate intervals*, and those limits are different for the ears of different people, and in any case they depend on the musical occasion and on the definition of the notes of the instrument concerned.

Modern Science's final observation is to point out to the 'theoretician' that he has now abandoned arithmetic; that he is relying instead on something quite different, on experimental observations of vibrations made by human ears; and that he would be 100 years behind the times, today, if he assumed that the ear is a perfect instrument for making such observations. Moreover, once he has admitted the ear into his picture he can no longer ignore it whenever it suits his convenience. He must follow it wherever it leads him. If he could begin all over again, today, he would find out all we now know about the way in which the ear reacts to musical vibrations and what we are learning about aural perception. In particular, he would

[7] See chapter 1, p. 13.

discover that while the musical ear attaches importance to what it accepts as the just intonation of the intervals of a concord on the accented beat of the music, it is singularly adaptable to a small change in the pitch of a note which may be called for by a succeeding concord. He might quite properly *begin* his investigations with experiments on the vibrations which his ear accepts as producing musical intervals, but he would not *end* there. He would find out why the ear's response to the perfect concords, the octave and the fifth, causes them to produce a *flexible* scale-system. In short he would learn to distil the musical scale from music itself, as imagined by the composer and heard by the musician through the co-operation of his ear and his brain.

5

THE HISTORY OF OUR SCALE

Scales are made in the process of endeavouring to make music and continue to be altered and modified, generation after generation, even till the art has arrived at a high degree of maturity.

SIR HUBERT PARRY, *The Art of Music.*

1. SCALES USED IN THE MIDDLE AGES

As Sir Percy Buck (1871–1947), himself the author of *Acoustics for Musician's*, has reminded us:[1]

> Music came first; then the scales accrued after ages of experiment; then came the theorists to explain them. And as they knew more of mathematics than of musical history they laid down laws which, in actual fact, no human being had ever obeyed.

This is a clear warning that to explain the scales of music is the task of musical scholarship. It is not an exercise in physics. The most the physicist can do to help is to undertake the calculations of musical intervals, perhaps using logarithms or their equivalents, which measure them. But the application of the results depends on the use made of these intervals in music. So to understand the origins of our musical scales we must first make some acquaintance with the early stages of English music.

Before polyphony developed there was a stage in which all music was melodic. The Church made use of plainsong, and in plainsong can be found the scale of Guido (990–1050). In his famous book *A Plaine and Easie Introduction to Practicall Musicke* (1597), a great English musical scholar, Thomas Morley, wrote out his instruction to Philomathes in the form

[1] *Proc. Mus. Assoc.*, vol. LXVI, p. 51.

of a dialogue between Master and Pupil. The first lesson begins 'Here is the Scale of Music which we term the Gam'. And then Morley draws a diagram of the *gam* in which we see the 'notes' of the scale arranged on a music stave, with the G, C and F clefs marked on the appropriate lines (Fig. 1).

In the annotations at the end of the book there is this note:

> That which we call the Scale of Music or the Gam, others call the Scale of Guido, for Guido Aretinus, a monk of the Order of St. Benet or Benedict, about the year of our Lord 960, changed the Greek scale (which consisted only of fifteen keys, beginning at A re (A) and ending at A la mi re (*a'*), thinking it a thing too tedious to say such long words as Proslambenomenos, Hypate hypaton,[2] and such like, and turned them into A re, B mi, C fa ut, etc.; . . .). But to the end that every one might know from whence he had the art he set this Greek letter *Γ* Gamma to the beginning of his scale, serving for a diapason [octave] to his seventh letter G; and whereas before him the whole scale consisted of four tetrachorda or fourths . . . he added a fifth tetrachordon, including the scale (but not with such art and reason as the Greeks did) seven hexachords or deductions of his six notes, causing that which before contained but fifteen notes contain twenty, and so fill up the reach of most voices. . . .

The annotations also contain a diagram of the diatonic *genus* of the Greek scale, as known in theory to Morley, in which the appropriate string-length as measured on a monochord is given for each of the 15 notes. For the Greeks discovered how the ratios of string-lengths may be used to verify or standardise musical intervals. Pythagoras (*c.* 570–500 B.C.) used a monochord, and under it he slipped a wedge-shaped bridge. He found that when one part of the string so divided was twice the length of the other part, they sounded notes a *diapason* (octave) apart, and when one was two-thirds of the length of the other they sounded notes a *diapente* (fifth) apart. Similarly when one part was three-quarters of the length of the other they sounded notes a *diatessaron* (fourth) apart. These Greek names for intervals were shown by Morley in his diagram of the Greek scale: so they are explained here to assist anyone who may have the opportunity to consult Morley's book. Applying Pythagoras' discoveries, and taking the string-lengths of a fifth and a fourth, we obtain the ratio of a

[2] The names of the lowest two notes of Morley's Greek Scale.

		[5]				
Double or Treble Keys[1]	ee	e''	La		La	1 note[7]
	dd	d''	La Sol		Sol La	2 notes
	cc	c''	Sol Fa		Fa Sol	2 notes
	bb	b'	Fa ♮Mi		Mi Fa	2 notes, 2 clefs[4]
	aa	a'	La Mi Re		Re Mi La	3 notes
	g	g'	Sol Re Ut	(Septima ut prima)	Ut Re Sol [G clef]	3 notes
	f	f'	Fa Ut	(Sexta ut tertia)	Ut Fa	2 notes
	e	e'	La Mi		Mi La	2 notes
Mean Keys	d	d'	La Sol Re		Re Sol La	3 notes
	c	c'	Sol Fa Ut	(Quinta ut secunda)	Ut Fa Sol [C clef]	3 notes
	b	b	Fa ♮Mi[6]		Mi Fa	2 notes, 2 clefs
	a	a	La Mi Re		Re Mi La	3 notes
	G	g	Sol Re Ut	(Quarta ut prima)	Ut Re Sol	3 notes
Grave or Bass Keys	F	f	Fa Ut	(Tertia deductio)	Ut Fa [F clef]	2 notes
	E	e	La Mi		Mi La	2 notes
	D	d	Sol Re		Re Sol	2 notes
	C	c	Fa Ut	(Secunda deductio)	Ut Fa	2 notes
	[2]♮	B	Mi		Mi	1 note
	A	A	Re		Re	1 note
	Γ	G	Ut	(Prima sex vocum deductio[3])	Ut	1 note

Fig. 1

[1] Keys = notes. In some early theoretical works it means clefs. [2] The B can only be ♮ because Mi was always a major 3rd above Ut; it could only be Fa (B♭) if there was an F Ut below the G Ut (but see p. 16). [3] 'Prima sex vocum deductio' = the first note from which six others are deduced, etc. 'Quarta ut prima' = the fourth (deduction) as the first (deduction), etc. [4] See p. 12. [5] Helmholtz's notation; if in the text, brackets will be used, *e.g.* C sol fa (c''). [6] M. has '♮ Mi,' but 2nd edition has '♮ Mi,' which is correct (b square). [7] Note = sol-fa name.

The footnote and page numbers on this diagram refer to Morley, *A Plain and Easy Guide to Practical Music* (*see* p. xii).

whole tone which is their difference. It then becomes clear from Morley's diagram that he is describing a scale used to tune the lyre that, as he says, consisted of *semitonium minus*, *i.e.* the *limma* of Pythagoras or the Greek hemitone, followed by *tonus* and *tonus*. These three intervals made up the tetrachord. This is Pythagoras' tuning by perfect fifths employed by the Greeks. In our modern terminology these three intervals would be a semitone less a comma, a major tone, and a major tone, which add up to a perfect fourth.

But there is, in Greek writings, in particular those of Claudius Ptolemeus (*c.* A.D. 90–168), mention of many tunings of the lyre; and we have also to take account of the intonation of the Greek aulos, a primitive form of oboe of much earlier date, that had equidistant finger-holes all of the same size. These

Fig. 2

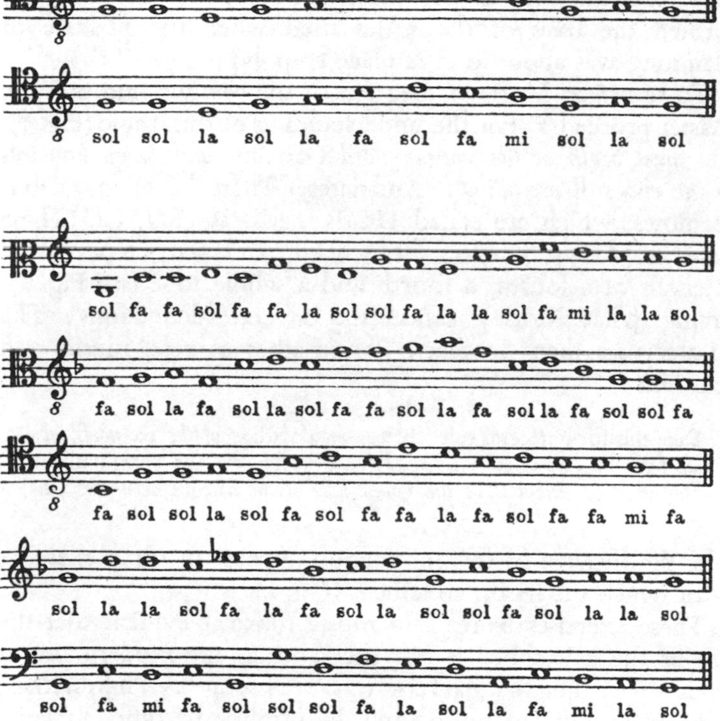

different tunings range from the mode of Olympus (*c.* 660–620 B.C.) to Didymus' (*c.* 63 B.C.–A.D. 10) diatonic, listed by Ptolemy, which resembles our modern major mode. As these covered a period of 700 years or more, a longer period than that from Dunstable to Vaughan Williams and Walton, it is obviously a mistake to write of Greek music as though it were all of the same kind, as some writers have done. It is inconceivable that music underwent no alteration or modification in all that time. The study of Greek music is the task of scholarship, in which problems remain to be solved. We shall therefore not attempt to discuss the Greek modes and *genera* here, especially as there has been uncertainty about them and text-books have been generally misleading. Moreover it is clear that these scales differed from the ecclesiastical modes derived from them, which found expression in plainsong. We shall therefore go no farther back into the past than to the scale of Guido (tenth century), described by Morley, that formed the basis of the scales used when the practice of plainsong was about to give place to polyphony.

To return to Morley's diagram of the *gam* we find that the Master proceeds: 'For the understanding of this Table (Fig. 1), *you must begin at the lowest word* Gam-ut, *and so go upwards to the end still ascending*'. And later, 'There be in music but six notes, which are called Ut, Re, Mi, Fa, Sol, La.' These notes made up Guido's hexachord, and they repeated at intervals of a fourth, a fourth and a whole tone (see Fig. 1), giving what Morley called the several 'deductions'. The Master uses them to teach Philomathes to read music, and tells him:

> Take this for a general rule, that *in one deduction of the six notes, you can have one name but once used*... but this we use commonly in singing, *that except it be the lowest note of the part we never use Ut.*

The significance of this instruction appears in some exercises with which the lesson concludes (Fig. 2):

These exercises in reading music make it evident that the use of the hexachord is not the same as the modern use of tonic *sol-fa*. The first exercise was 'well song' by Philomathes; but, as the staff notation shows, he produced a different series of notes from that produced in tonic *sol-fa* by a singer who sang *soh soh lah soh lah fah soh fah me soh lah soh*, perhaps using the modulator. Only the first two, the fourth and the fifth, and the last three notes would be the same. But the two systems have one thing in common: both are methods of teaching sight reading in singing.

The names of the twenty notes of the *gam*, as given in Fig. 1, were the first pitch notation given to these notes, and they remained in use till the eighteenth century. Thus the Ds in successive octaves were called D *sol re*, d *la sol re*, and dd *la sol*.

There is reason to think that the scale that was used for plainsong actually produced Pythagorean intonation, *i.e.* that obtained through the Greek tuning by perfect fifths, or something like it. This would be a possible melodic scale and could readily be obtained by means of a monochord such as Boethius (*c.* 480–524) used to define the Pythagorean scale. Boethius was one of the 'Ancient Writers' cited by Morley, whose books transmitted knowledge of the musical art of the Greeks to the middle ages and later.

2. The Scales of Early Polyphony

Let us now turn to some early music for singing in parts; and because it is still sung, and is well known to day, let us take as our earliest example the thirteenth century round 'Sumer is icumen in' which is here, Fig. 3, transcribed from *Grove*.

Of this composition Dom Anselm Hughes, O.S.B., writes in *Grove:*

> In six directions is it pre-eminent, for (i) it is the oldest known canon; (ii) it is the oldest known harmonised music which is frequently performed and enjoyed by singers and listeners today; (iii) it is the oldest known 6-part composition; (iv) it is one of the oldest known specimens of the use of what is now the major mode; (v) it is the oldest known specimen of ground bass; (vi) it is the oldest known manuscript in which both secular and sacred words are written to the music.

It is because it is the oldest known harmonised music that is frequently performed to day and is written in what is now the major mode that it has been chosen for the earliest example of polyphony. The mode was described as the Ionian mode *transposed*, which means that it was transposed down a fifth.

So instead of appearing to us to be written in C major it seems to be written in F major.

Fig. 3

Cantus i

Cantus ii

Cantus iii

Cantus iv

Bassus ii

Bassus i

etc.

The harmony, consisting of concords with occasional passing notes, is extremely simple. The harmonic outline is:

repeating every four bars. How far the singers would succeed in singing truly in tune, *i.e.* in producing true intonation between the parts, and how far they would be influenced by the melodic scale of plainsong, to which they were accustomed, we cannot pronounce authoritatively today. Walter Odington (thirteenth century), however, states that singers intuitively used true intervals instead of those given by the Pythagorean monochord. But as polyphony was increasingly used singers became increasingly accustomed to singing in parts, and skilled singers would certainly produce true intonation.

So let us consider the intonation that would be produced in 'Sumer is icumen in' by such singers. The intonation they would use in the first, third, and fourth bars calls for no comment. But consider the intonation of the second bar. Here we have a minor triad on what we should call the second-degree note of the scale. This triad is comprised between the second- and the sixth-degree notes of the scale, which would be so sung as to produce a true fifth.

Fig. 4

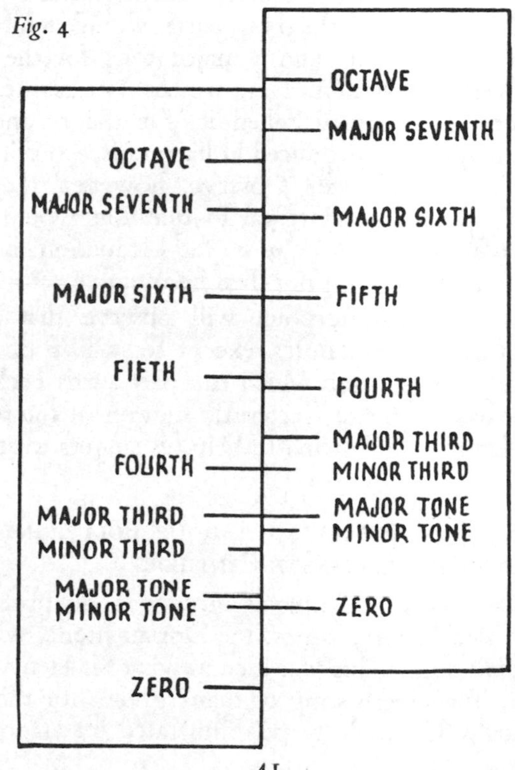

Sir Charles Stanford has distilled from sixteenth-century polyphony the intonation produced by singers from these notes. In Fig. 4 there is a diagram that explains his conclusions. The right-hand side shows the intervals of a minor triad consisting of a minor third and a fifth measured from the same 'zero'. The left-hand side shows the intonation that this triad produces in the following notes of the Ionian mode *transposed*, reckoning from the final (F) as zero, *viz.*:—the major sixth, the fourth, and the whole tone. Using our modern names for them, these three notes would be D, B♭ and G. The diagram makes it clear that the note a perfect fifth below D is a minor tone above the zero F. This is important, for from it we learn that in the second bar the tendency of *Bassus* i would have been to sing the interval of a minor tone between F and G. Remember that this seems to our modern ears to be the scale of F major, so that the intonation of the interval we are concerned with is that between F (*ut*) and G (*re*). The singers of the bass parts would tend to use, alternately, a minor tone and a major tone for the interval FG in the harmony written. Here we have an example of the mutable note, as Stanford called it, for the second-degree note of the major scale produced in harmonised music by true intonation in performance. Observe, however, that in the Ionian mode proper the interval in question would be that between F (*fa*) and G (*sol*), *i.e.* in the 'deduction' beginning with C as *ut*, and G would not then be mutable.

The student of counterpoint will observe that the bass parts consist of conjunct notes, except for a leap of a minor third, from A to C, after which the part turns back to B♭. He will know that this characteristic motion of the parts was employed, later, by Palestrina to help his singers to sing truly in tune.

3. The Scales of Folk-song and of Polyphony in the Fifteenth and Sixteenth Centuries

Let us now take an example of another mode, probably the most often used in early times: the Dorian mode, which was mode i of plainsong. This was used in what Sir Henry Hadow described as 'the superb song of thanksgiving for the victory of Agincourt'. The melody was illustrated by Hadow in his

English Music. As an example of other music in the Dorian mode consider the folk-song 'Greensleeves' mentioned by Shakespeare in *The Merry Wives of Windsor* (Fig. 5):

Fig. 5

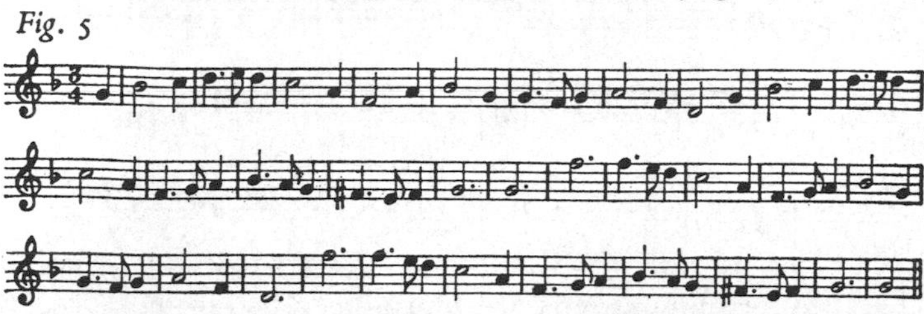

Observe that this melody, as written out in Fig. 5, is in the Dorian mode *transposed* down a fifth, as shown by the B♭ in what we should now call the key signature. This example has been chosen because its mode is that of the very famous 'Western Wynde' Mass of Taverner (*c.* 1525) which was also written in the Dorian mode *transposed*, though the character of the mode was not affected by the transposition.

We may form an idea of the Dorian mode by thinking of a scale on the pianoforte, using white notes only, and running up from D through an octave:

But this is an oversimplification, for the scale system of the modes was hexachordal. This was a relic of plainsong and it made use of the six notes of Guido's hexachord:

ut, re, mi, fa, sol, la

As our examples are in the Dorian mode *transposed* we will take the mode in that form. The hexachord in the Dorian mode *transposed* was:

G, A, B♭, C, D, E, with, on occasion, E♭ used to avoid the tritone, and with an extension upwards to F and an extension downwards to F♯.

These nine notes were all definite notes of the scale, each with its own proper function. F and F♯ were not alternatives, as in our scale to day, but equally important and independent notes. So were E♭ and E. As the melody 'Greensleeves' shows, all the notes except E♭ occur therein and they lie within an octave (*see also* Chapter 8, p. 71).

Fig. 6

Now let us turn to Taverner's 'Western Wynde' Mass of which the opening strains are reproduced in Fig. 6. Observe that these strains of the Mass consist almost entirely of concords that, if it is listening to a good choir, the musical ear will expect to hear sung in tune. Anyone who plays it with a soft stop on the organ (or with the *voix celeste*, which disguises to some extent the equal temperament to which our modern organs are tuned) will be impressed by its harmonious quality and by the modal character of the music. Like 'Greensleeves', it seems to our modern ears to alternate between the keys of G minor and F major, as a rule without modulations that 'establish' either key, and occasionally it seems to touch the relative major and minor of these two keys. As one accustoms oneself to modal compositions, such as this, one begins to recover, however imperfectly, the modal sense of music composed before 'tonality' as we know it asserted the importance it acquired later.

Careful examination of the harmony suggests that at least three of the notes of the hexachordal scale of the Dorian mode *transposed* have to be mutable—G, A and C. We have already found that in the Ionian mode *transposed*, which resembles our key of F major, G, which seems to us to be the second-degree note, when harmonised as part of the prevailing harmony has to be mutable, on occasion, so as to sound perfectly in tune. This is what Stanford tells us, and he also explains that in a minor scale the second- and seventh-degree notes have to be mutable when required,[3] while the fourth-degree note of a minor scale is lower than the second-degree

[3] C. V. Stanford, *Musical Composition*, see pp. 15 and 16. Mutable notes are also explained in the author's *Music and Sound*, see pp. 13, 14, 15 and 16; also in his *Musical Slide-Rule* [Ref. 3], pp. 8, 9 and 10. See also Fig. D.

note of the relative major.[4] Thus when the 'Western Wynde' Mass sounds like F major to our modern ears we should expect G to be mutable on occasion, and when it sounds like G minor we should expect A, and possibly F, to be mutable, and certainly we should expect C to be a note somewhat lower than the C we hear when the music sounds like B♭ major. The modal scale of the Dorian mode was surprisingly flexible.

Mutable notes are produced by concords sung with true intonation; they therefore cause no difficulty of intonation. All that is necessary is that they should be sung or played perfectly in tune[5] and they will then find their own positions. But they presented a problem to the tuner of a keyboard instrument. Readers should be warned that in physics text-books which attempt to describe scales no mention is made, as a rule, of mutable notes. Yet they are the main reason why the scales of music are flexible in intonation, from which arises the need for tempered intervals that physics text-books accept for keyboard instruments with their rigidly fixed intonation.

In the two examples of modal music that have been discussed, 'Sumer is icumen in' and the opening strains of Taverner's 'Western Wynde' Mass, most of the notes were found to be elements of concords that formed part of the prevailing harmony. These the musical ear would wish to hear sung in tune, giving effect where necessary to the mutable notes. But a few of the notes were passing notes and their intonation (a word that implies performance) is not controlled in the same way by the prevailing harmony. The ear of the musician would find the melodic outline of each part to be the important thing, and its intonation would depend on the sensitiveness of the singer's musical feeling. While he is singing passing notes the singer is looking ahead, and his feeling is for the essential note aimed at; in the same way as the music student writing, in an exercise, third species counterpoint, or four

[4] See also *Music and Sound* [Ref. 1], p. 14, Fig. 6, which exhibits graphically this difference.

[5] If intervals are not perfectly in tune there is roughness due to the sensation of beating, as Helmholtz explained. The singer intuitively avoids this sensation. For an account of Holmholtz' explanation see the author's *Music and Sound* [Ref. 1], Chapter V, or *The Musical Ear* [Ref. 11], pp. 21 *et seq.*

notes against one, looks ahead to the concord aimed at. This is true of all decorating notes sung or played on strings. That is why the ear finds such satisfaction in unaccompanied part singing or in a string quartet.

By the end of the sixteenth century the modes were ceasing to be used, especially in England. The successive notes of the major scale formed, with the keynote, intervals that were either major or perfect, *e.g.* major third, perfect fourth, perfect fifth, major sixth, and major seventh. To this general observation the second-degree note might be an exception for it is a mutable note. Fig. 6 on p. 14 of the writer's *Music and Sound* (already referred to in a footnote) shows how two of the notes of the major scale are displaced in the relative minor scale.

4. The Scales of the Seventeenth, Eighteenth and Early Nineteenth Centuries

Sir Donald Tovey has observed:[6]

> The chief and not wholly unconscious aim of the successors of Monteverdi (that is to say, of the composers of the mid-seventeenth century) was to establish the tonic-dominant-subdominant orientation of major and minor keys in a system which could digest essential discords. A modal composition visited other modes than its own whenever it made a cadence other than on its own final; but it did not establish itself in the visited modes; and still less did it go into regions that produced its own mode at a different pitch.

At the beginning of this chapter a passage was quoted in which Sir Percy Buck warned us against theorists who 'know more of mathematics than of musical history'.[7] We draw the same warning from the passage just quoted from Sir Donald Tovey. One recalls more than one book on physics that discusses the development of scales, and observes that at first composers could give variety to a musical composition by writing in many different modes. That is to confuse *modes* with *keys*. The reader will observe that Tovey writes, not of 'many different modes' which could be used in a composition, but of a composition as having 'its own mode'. But the seventeenth

[6] *Encyclopaedia Britannica*, 14th ed., article 'Harmony'.

[7] Doubtless, to judge from this description, he had particularly in mind A. J. Ellis, the translator of Helmholtz' *Tonempfindungen*, to whom his warning certainly applies.

century provided a contrast: a composer could then use many different keys in a composition if he so wished, while the modes had become reduced to two, major and minor. In this way the scale was modified; for although each key used the same scale in the major or minor mode as the case might be, each used the scale at a different pitch. The essential thing is that the system of key-relations, and modulation, must be understood; and they are a part of musical history. That is why the student is given exercises in modulation to work at an early stage in his harmony lessons.

We have seen that the chief characteristic of the scales that can be distilled from modal polyphony is their flexibility. We shall therefore confine ourselves to discovering how the composers of the seventeenth and subsequent centuries increased the flexibility of the musical scale when they introduced chromatic writing and modulation using, it may, be a considerable variety of *keys* in the course of an extended musical composition. Stanford described the mutable notes of the major and minor scales. Each new key used at a different pitch would make more notes mutable.

We have also learnt that passing notes would increase the flexibility of the scale in modal music. They would have the same effect in later music in which practice allowed the use of as many keys as the composer wished to employ. The intonation of the passing notes, especially if the music moved at all rapidly, would be somewhat indeterminate because they did not form part of the prevailing harmony.

A further increase in the flexibility of the scale was brought about by enharmonic change, by which is meant the substitution of, say, G♯ for A♭, in a modulation. Now on our modern keyboard-instruments the difference between these two notes is disguised by equal temperament, and both are played by the same black key. We learn also that when organs were tuned in mean-tone temperament there were organs, *e.g.* that in the Temple Church, which produced separate sounds for G♯ and A♭, and an enharmonic change from one note to the other would involve an actual change of finger from the sharp to the flat, or *vice versa*. Thus there would be an actual break in the tone. Perhaps that is why the possibilities of enharmonic change were supposed to be the special virtue of equal temperament (as compared with mean-tone temperament), because there would be no break in the tone as the same black key would be used for both notes. Sir Donald Tovey[8] writes:

> The object of 'enharmonic' modulation is frankly to mystify. It is popularly supposed to belong specially to tempered scales; but it really presupposes just [*i.e.* true] intonation. . . . Even with a limited keyboard the ear imagines a change of intonation when the unexpected resolution appears.

But with the strings we are not dependent on what aural perception can achieve in this way; the change from one chromatic note to the other is actually made without any break by the player. His musical sense, responding to the change of harmony, impels him to make, it may be intuitively, the small change in the point of application of the pressure of his finger-tip required to effect the small change of note. So then a chord containing say G♯ dissolves mysteriously into one containing A♭. Thus enharmonic change adds to the flexibility of the musical scale, and it did so particularly in the music of Beethoven and Wagner.[9]

Then there is the violinist's practice of closing the leading note so as to bring it nearer to the tonic in a melodic line when there is no dominant sounded with it to control its intonation. There is an interesting passage in Moritz Hauptmann's *Letters of a Leipsic Cantor*, here taken from a translation in Pole's *Philosophy of Music:*

> In the two passages C, D♭, C and C, C♯, D, it is certain that the D♭ will be sung flatter than the C♯. . . . This is the psychological view of intonation that the clavier can know nothing of.

It gives point to this passage that theory tells us that C♯ is lower than D♭, for C♯D and CD♭ are diatonic semitones, and these intervals are each more than half a whole tone. They will therefore have to overlap if they are thought of as filling up, between them, the whole tone CD, which will make C♯ lower than D♭. The quoted passage means that in the experience of the distinguished choir-master and violinist,

[8] *Encyclopaedia Britannica*, 14th ed., article 'Harmony'.

[9] It is recorded that Wagner's use of enharmonic change in *Tristan* was particularly objected to by Berlioz, whose own use of enharmonic modulation was different from Wagner's, as Saint Saëns pointed out (see: *The Heritage of Music*, vol. II, p. 141).

Hauptmann (1792–1868), it is the melodic character of decorating notes, which do not form part of the prevailing harmony, that determines their intonation. This again increased the flexibility of the scale during the 250 years that followed the golden age of polyphony. This has its effect on the choice of temperaments. But we must not suppose that the notes we hear (perceive) always correspond quite exactly to the keyboard sounds we listen to.[10] On one occasion, in discussing the intonation of the pianoforte, Professor Dent[11] took the progressions C, D♭, C and C, C♯, D, accompanied by suitable harmonies, which he played, to illustrate the fact that the same black note on the pianoforte might have slightly different effects on the ear in differing musical circumstances, and be *perceived* as having different intonations.

5. Scales Developed by Particular Composers in the Nineteenth Century and After

A series of notes that some composers have found gave intervals they could use for special effects is that known as the whole-tone scale: C, D, E, F♯, G♯, A♯ (= B♭), C, in which there may be an enharmonic change somewhere, say G♯ to A♭ or A♯ to B♭. This gives a scale of six whole tones, which in some composers' music are whole tones of equal temperament. It depends on whether the pianoforte or an instrument of free intonation is used whether these whole tones are equal-tempered tones or are the product of enharmonic change.

Sir Charles Stanford observes:[12]

> The student of some interesting modern developments will also speedily discover that the adoption of the so-called whole-tone scale as a basis of music is, except upon a keyed instrument tuned to the compromise of equal temperament, unnatural and impossible. No player upon a stringed instrument can play the scale of whole tones and arrive at an octave which is in tune with the starting note, unless he deliberately changes one of the notes on the road, and alters it while playing it.

which is another way of stating what was written above about enharmonic change in the tonal scale.

[10] This is discussed in the final essay of the author's *Musical Ear*. [Ref. 11].
[11] *Proc. Mus. Ass.*, vol. LXXVII, p. 50.
[12] C. V. Stanford: *Musical Composition*, p. 17.

The tonal scale turns up again and again in the later Russian music, in particular in the compositions of Moussorgsky (1839–1881), which in turn suggested it to certain French composers, particularly Debussy (1862–1918). But on this Edwin Evans has observed [13] 'the tonal scale is not Debussy, nor is Debussy the tonal scale'. Professor Shera writes of the tonal scale[14] 'it is more than probable that we ought to regard it as a chord rather than a scale', and he proceeds later to consider some of its 'very interesting chordal properties'. Sir Donald Tovey observes[15] 'It is really no more a whole-tone scale than the diminished seventh is a major sixth bounding a series of minor thirds'. To illustrate his meaning he adds that Sir Walford Davies 'has pointed out that this scale is a six-note chord projected into a single octave and capable, like the diminished seventh, of an enharmonic turn to each of its notes'. It lent itself readily to use in some pianoforte compositions and songs with pianoforte accompaniment by Debussy, where its elusive properties were useful in suggesting impressionistic effects; but its possibilities were soon exhaused, and few composers are likely to attempt new uses of the tonal scale.

Some writers of physics books, attracted by the mathematical possibility of dividing an octave into six equal intervals, which they can measure with their logarithms, have described this so-called scale as though it were of the same nature as the diatonic scale. They failed to realise that, as Martin Cooper has observed,[16] 'There is a great difference between the fairly frequent use of whole-tone chords, or even whole passages, such as can be found in Liszt (1811–1886), Saint-Saëns (1835–1921), and Rimsky-Korsakov (1844–1908), and the conscious adoption of the whole-tone scale as an alternative to the hitherto accepted diatonic order'. The observations of these physicists would lead one to expect that the tonal scale is *often* found in Debussy which, for the reasons given, would

[13] *Proc. Mus. Ass.*, vol. XXXVI, p. 66.

[14] In his 'Musical Pilgrim' volume *Debussy and Ravel*.

[15] *Encyclopaedia Britannica*, 14th ed., article 'Harmony', where the academic possibilities of the enharmonic resolution of the chord are shown on a music stave.

[16] *French Music*, p. 41.

be a considerable overstatement. Certainly its chordal possibilities were fully explored by him in some pianoforte compositions, but without leaving much for anyone else to discover.

Finally, a word or two must be added about the twelve-note technique adopted by Schoenberg and others. Opinion amongst musicians appears to be crystallising in two forms: on the one hand that of Schoenberg's ardent admirers, and on the other those who think that, like other experiments that derive from equal temperament, twelve-note music is leading to a *cul-de-sac*. That question will be finally decided by the musicians of the next generation. Those who wish to understand this and similar tendencies in modern music are recommended to read Edwin Evans' article 'Atonality and Polytonality' in Cobbett's *Cyclopedic Survey of Chamber Music.*

6

TEMPERAMENTS—WITHOUT TEARS

THE method of tuning the piano with which we are familiar is an arithmetical puzzle to many musicians, because the text-books which explain it generally expect them to understand the twelfth root of 2. If they nevertheless pursue the matter they probably find that the author goes on to use logarithms, and in the last paragraph of this article we shall see that he has good reason on his side. But if logarithms are a mystery to them, their courage may fail them, and temperaments will remain an unsolved puzzle.

That is a pity, for the older theorists, like Zarlino (1517–1590), discovered pretty well all there is to know about temperaments, though they did not use logarithms which had still to be invented in their day. But they could use ordinary arithmetic, so there must be some way of explaining temperaments by simple arithmetic. This article offers an explanation which could be understood by a schoolboy—not Macaulay's schoolboy, but one in the lower fourth who has mastered addition and subtraction in algebra, which deal with numbers disguised as letters and are really simple arithmetic.

Let us begin by contradicting some text-books which tell us the temperaments became necessary only when music began to make use of modulation as we know it. That is not true. Temperaments are necessary on any keyboard instrument because all its notes are fixed in pitch, and there are only twelve of them in each octave. They may or may not make it possible to modulate freely. Equal temperament makes it possible to modulate freely, but only because it mistunes all the intervals except the octave—the thirds and sixths rather drastically. On the other hand, mean-tone temperament, invented in the fourteenth century and used in this country till about the middle of last century, permitted only a limited degree of modulation because it made all the major thirds true—or sweet as the tuner called them. The five black keys in each octave were tuned as C♯, E♭, F♯, G♯ and B♭; so there was no D♭, D♯, G♭, A♭, or A♯.

The fact is that if we tune upwards from a given note in a series of octaves, perfect fifths, or true major thirds, we never reach a unison again between the notes of any two of the series, however far we go. So when we attempt to make fifths or thirds fit into octaves there are bound to be misfits. The older theorists knew all about the small discrepancies that result, and they had names for them.

Thus they said that four perfect fifths exceeded two octaves and a major third by a small interval called a *comma* (*of Didymus*). Eight perfect fifths and a major third exceeded five octaves by a still smaller interval called a *schisma*. Twelve perfect fifths exceeded seven octaves by an interval called the *comma of Pythagoras*, because the Greeks discovered it. One octave exceeded three true major thirds by an interval called a *diesis* (see also pp. 177, 217).

Now try these combinations of octaves, fifths and thirds on the piano. You will find that four fifths are exactly two octaves and a major third, that eight fifths and a major third are exactly five octaves, that twelve fifths are exactly seven octaves, and that three major thirds fill an octave exactly. All the discrepancies have disappeared. What does this mean?

With the schoolboy's help we can easily find out what has happened to the fifths and major thirds, assuming the octaves to be true. Call the comma k, the schisma s, the comma of Pythagoras p and the diesis d; and for the perfect fifth write V (only as shorthand), for the octave write VIII, for the true major third write III. We can now represent our discrepancies by equations if we think of plus as tuning an interval up, and minus as tuning an interval in the opposite direction:

$$4\text{V} - 2\text{VIII} - \text{III} = k \quad \ldots\ldots \quad (1)$$

$$8\text{V} + \text{III} - 5\text{VIII} = s \quad \ldots\ldots \quad (2)$$

$$12\text{V} - 7\text{VIII} = p \quad \ldots\ldots \quad (3)$$
$$\text{VIII} - 3\text{III} = d \quad \ldots\ldots \quad (4)$$

In these four equations there is hidden the key to all practicable temperaments, *i.e.* to all means of tuning a keyboard instrument, with seven white and five black keys to each octave, so as to make it playable.

In any temperament, what we have to do is to find a way of getting rid of the discrepancy in equation (1) which is k. Add equation (1) and equation (2) together. We find that:

$$12\text{V} - 7\text{VIII} = k + s$$

Hence it follows from equation (3) that:

$$p = k + s \quad \ldots\ldots \quad (5)$$

Now multiply equation (1) by two, when it becomes:

$$8\text{V} - 4\text{VIII} - 2\text{III} = 2k$$

Subtract from it equation (2). The schoolboy tells us that to subtract —5VIII we change the minus sign into + and add. Or as musicians we can say: each minus sign changes the direction of tuning, so two minus signs mean 'tune up'. So making our subtraction we find that:

$$\text{VIII} - 3\text{III} = 2k - s$$

Hence it follows from equation (4) that:

$$d = 2k - s \quad \ldots\ldots \quad (6)$$

Equations (5) and (6) tell us what we want to know about equal temperament without worrying about the twelfth root of 2.

First, however, let us go back a century or more, to meantone tuning. The objective of this temperament was to make all major thirds quite true; and of course the octave was made true. So supposing our piano had an extra note for A♭, we should find that A♭C + CE + EG♯ would be a diesis short of an octave (see equation (4)). Thus on any piano in which octaves and major thirds were tuned perfectly true the interval G♯A♭ would be a diesis. And this is quite a considerable interval, so considerable that on the organ which Father Smith built for the Temple Church in 1684 he divided the black key for G♯ across the middle and made the back half, which he raised, play A♭. But on organs without this device, if G♯ was played when A♭ was wanted there was an unpleasant dissonance, which was called a wolf because of its howling. This wolf, and others like it—produced every time C♯ or E♭ was played when D♭ or D♯ was wanted, and so on—were the price mean-tone tuning paid for its sweet thirds. The number of good keys was limited, and modulation was very restricted.

How about the fifths? Here we go back to equation (1) and see what happens if we get rid of the discrepancy by dividing the comma among the four fifths. Thus:

$$4(\mathrm{V} - \frac{k}{4}) - 2\mathrm{VIII} - \mathrm{III} = 0$$

All octaves are tuned true, so in this temperament all the fifths were made too narrow by a quarter of a comma.

But the Greeks avoided the discrepancy in a different way. For their diatonic genus they used a third which added a comma to our major third, thus:

$$4\mathrm{V} - 2\mathrm{VIII} - (\mathrm{III} + k) = 0$$

This made the fifths true, while the third (the Pythagorean third) was much too wide for use, some centuries later, in harmony.

In the eighteenth century a compromise was adopted by Silbermann[1] (1683–1753), the famous organ builder of Bach's day. He both flattened the fifths and sharpened the thirds. He made the fifth only one-sixth of a comma flat, so that in his tuning, if the third had remained true, we should have:

$$4(\mathrm{V} - \frac{k}{6}) - 2\mathrm{VIII} - \mathrm{III} = k - \frac{2k}{3} = \frac{1k}{3}$$

and there would still be a small discrepancy left. To get rid of this it was added to the major third and our equation became:

$$4(\mathrm{V} - \frac{k}{6}) - 2\mathrm{VIII} - (\mathrm{III} + \frac{k}{3}) = 0$$

As the discrepancy has disappeared this temperament could be realised on a keyboard with seven white and five black

[1] J. Murray Barbour, *Musical Quarterly*, January, 1947, 'Bach and the Art of Temperament'.

keys to the octave. The wider (tempered) thirds made the tempered diesis less than half the size of a true diesis, so the wolves howled much less, and more keys could be used with little ill effect. Only the remote keys remained impossible. The fifths were improved as compared with mean-tone fifths, but the thirds suffered. This tuning was a halfway house to our modern equal temperament.

Remembering that $p = k + s$, we may try another compromise.

We can write equation (4) thus, by subtracting $\frac{p}{3}$ (*i.e.* $\frac{4p}{12}$) from both sides:

$$4(\mathrm{V} - \frac{p}{12}) - 2\mathrm{VIII} - \mathrm{III} = k - \frac{p}{3}$$

$$= \frac{3k - k - s}{3}$$

$$= \frac{2k - s}{3}$$

$$= \frac{d}{3}$$

or:

$$4(\mathrm{V} - \frac{p}{12}) - 2\mathrm{VIII} - (\mathrm{III} + \frac{d}{3}) = 0$$

which again gets rid of the discrepancy.

This is how our pianos are tuned today. In this modern tuning (equal temperament) three tempered major thirds will be a whole diesis more than three true major thirds and therefore exactly fill an octave. The wolves will be silenced. But equation (3) becomes:

$$12(\mathrm{V} - \frac{p}{12}) - 7\mathrm{VIII} = 0$$

so that by making twelve fifths exactly equal to seven octaves we get rid of the diesis, which means that we can use one black key for each pair of neighbouring sharps and flats. That is why we can play up twelve fifths in succession on the piano, beginning with bottom A, and finish up on the top A, seven octaves higher. It was not possible to do this on a piano tuned in mean-tone temperament because G♯E♭ was the wolf-fifth and the succession of fifths broke down there.

The price paid for getting rid of the wolves altogether was no light one. All the thirds and sixths are about two-thirds of a comma too sharp or too flat. Indeed it is evident, from the various changes we have made in equation (1) to get rid of the discrepancy of a comma, that we can always improve the fifths at the cost of the major third and *vice versa.*

Equation (1) teaches us one more thing. Suppose we tune up two fifths from middle C. We reach D near the top of the treble stave. If we subtract an octave we are left with CD, a whole tone. Call it T. So:

$$2\mathrm{V} - \mathrm{VIII} = \mathrm{T}$$

Now tune up an octave and a major third from middle C. We reach E in the top space of the treble stave. Subtract our two fifths; we are left with DE, again a whole tone. But we must not assume that it is the same size as T, so call it t. Thus:

$$\mathrm{VIII} + \mathrm{III} - 2\mathrm{V} = \mathrm{t}$$

Subtract the second of these equations from the first (remember that to subtract $-2\mathrm{V}$ we write $+2\mathrm{V}$ and add). So we get:

$$4\mathrm{V} - 2\mathrm{VIII} - \mathrm{III} = \mathrm{T} - \mathrm{t}$$

Hence from equation (1) above:

$$\mathrm{T} - \mathrm{t} = k$$

So in all untempered music there are two sizes of whole tone, and one is a comma larger than the other.

When we get rid of the small discrepancy k, in tuning, we alter T and t until $\mathrm{T} - \mathrm{t} = 0$,

$$i.e.\ \mathrm{T} = \mathrm{t}$$

This means that in each of our temperaments all whole tones are the same size, and each one must be half a major third in that temperament. Consequently a tempered whole tone is least in mean-tone temperament, largest in equal temperament, and of intermediate size in Silbermann's temperament. In an octave there are five whole tones and two semitones. So mean-tone tuning has the largest semitone and equal temperament the smallest one.

One final observation. Using his logarithms the mathematician discovers that, as nearly as the ear can tell, there are eleven schismas in a comma, twelve in a comma of Pythagoras, and twenty-one in a diesis, while a fifth of a whole tone is nearly a diesis. Which gives an idea of the relative sizes of these small discrepancies.

[Note, that elsewhere in this book, III is denoted as M3, V as 5th, VIII as 8th, and *s* is used for the distance semitone—*not* the schisma. H.B.]

7

EQUAL TEMPERAMENT

ANYONE who looks through the fifth edition of Grove's *Dictionary of Music* and examines the many illustrations of keyboard instruments, from the portative organ (as used c. 1400) on Plate 47 to the modern pianos on Plate 54, both in Volume VI, will find that throughout this period of 550 years the keyboards shown there have one thing in common; there are always seven 'white' keys and five 'black' keys to the octave. Sometimes the 'white' keys were black and the 'black' ones were white; but that is merely a matter of fashion. A figure on p. 290, Vol. VI, shows that this arrangement of keys (7 + 5) was foreshadowed on the Halberstadt organ, built in 1361. Its keys were clumsy things intended to be thumped down by the hand. Their number was therefore not fixed by considerations of convenience for the fingers; fingered keys belong to a later date with improved mechanism.

Evidently, therefore, it was for some musical reason that five 'black' keys were considered to be enough for the Halberstadt organ. Such a reason we find in the fact that the lettering on the five 'black' keys shows that they gave all the chromatic alternatives for 'white' ones that were permitted by strict modal practice, *i.e.* B♭, E♭, F♯, C♯ and G♯. Indeed, when this organ was built five 'black' keys would be regarded as a generous supply. For, long afterwards, some keyboards had only one such key, B♭, required to sound *b molle* in the gam, or hexachordal scale, due to Guido, which was described by Thomas Morley[1]. On Plate 90 (or 101) of the *Oxford Com-*

[1] Morley's diagram of the gam is reproduced on p. 36, Chapter 5.

panion to Music there is a diagram, reproduced from Virdung's *Musica Getutscht* (1511), of an early keyboard with the added B♭. Such keyboards survived till Praetorius's time, and were described by him in his *Syntagma Musicum* (1619). The seven 'white' keys would give the remaining notes of the gam: *C fa ut*, *D sol re*, *E la mi*, *F fa ut*, *G sol re ut*, *a la mi re*, and *b mi* (now called B natural). Today we think of these notes as the seven notes of the diatonic scale of C.

In a sense it was a happy accident that, when the fingered keyboard came into use, this combination of five 'black' keys and seven 'white' ones in an octave was found to be so convenient for the human hand that musicians have never allowed it to be altered. That the keyboard as we know it is well adapted to the human hand is what made it inevitable that equal temperament should eventually replace mean-tone tuning when composers had begun to write music that made the wolves of mean-tone tuning quite intolerable. As soon as musicians became impatient with these wolves they set to work, first of all to make them howl less, and finally to silence them altogether. An example of the first stage was the tuning used by Silbermann, the great German organ-builder of J. S. Bach's day. The final stage was the universal adoption of equal temperament.

Theorists, obsessed with the arithmetical niceties of temperaments and the manifestations of the twelfth root of 2, often failed to realise that the keyboard, as we know it, imposed on all choice of temperaments an over-riding limitation from which there was no escape. Ever since Zarlino's day there have been those who have attempted to improve the musical resource of keyboard instruments by adding to the number of keys in the octave. These attempts have always failed to win permanent acceptance. As Robert Smith observed in his *Harmonics* (1759):

> The old expedient for introducing some of these sounds (D♯, A♯ . . . A♭, D♭, G♭, etc.), by inserting more keys in every octave is quite laid aside by reason of the difficulty in playing upon them.

It is often said that equal temperament has made a marked contribution to the art of music. That statement calls for some qualification of its literal meaning. The mistuned intervals of this temperament make no contribution to the art of music.

Good violin players are conscious of them on the pianoforte, and some pianists of sensitive hearing are well aware of differences between the intonation of a good string quartet and that of a pianoforte. A good orchestra does not play in equal temperament, nor must the faulty intonation of a poor one be identified with equal temperament. The real contribution that equal temperament has made to the art of music is indirect; it lies in the fact that, in spite of its mistuned intervals, equal temperament made possible the continued use of a keyboard with twelve notes to the octave, such as had been used for mean-tone tuning. To that continued use of a twelve-note keyboard is due the nineteenth-century development of pianoforte technique and pianoforte music.

The immediate purpose of the previous chapter was to provide an arithmetical explanation of temperaments without introducing the twelfth root of 2 or relying on calculations made by means of logarithms. From this explanation it emerged that any practicable temperament has to begin by eliminating from its intervals all differences of a comma, a comma being defined as the excess of four perfect fifths over two octaves and a true major third. It was found at the end of the chapter that tuning the diatonic scale of C with true intervals would produce two sizes of whole tone. The interval C to D would be a greater whole tone, and the interval D to E would be a lesser whole tone, and the two intervals would differ by a comma. Reasoning similar to that then used (p. 57, previous chapter) could make use of tunings up from F, instead of from C, and would show that, in the diatonic scale of C, G-A also is a lesser whole tone. Since C and G would be tuned a perfect fifth apart, the two notes a lesser whole tone above each of them respectively, would also be a perfect fifth apart. This means that to make D-F-A a true minor triad on the second-degree note of the scale we should need a new D, a lesser whole tone above C. But a twelve-note keyboard cannot produce pairs of notes a comma apart. So, when we say that any practicable temperament must begin by eliminating from its intervals all differences of a comma, the statement just amounts to this: that the temperament must be usable for tuning a twelve-note keyboard, the only practicable one.

In the same chapter it was found that in any such temperament the tuning of the fifths can be improved at the cost of the major thirds, and *vice versa.* So we must think of temperaments as being, essentially, *practical* ways of tuning a keyboard with twelve notes to the octave; and we must think of the eighteenth-century tuner as aiming at so much sharpening of the major thirds as he was prepared to allow in order to make the wolves howl less. Finally, for the occasional howling of a wolf, later tuners substituted a general mistuning of all the intervals and called it equal temperament.

To think of temperaments in this way is to understand their history. The sharp tuning of major thirds developed most rapidly in Germany. But there were tuners in the Latin countries, and even in Britain, who modified mean-tone tuning by some sharpening of the thirds. This was long before the advantages of equal temperament were generally admitted to outweigh its disadvantages, and it was the changing practice of composers that eventually turned the balance. To the more adventurous tuners the adoption of equal temperament meant going only one logical step farther than they had gone already. To conservative tuners, whose ears were accustomed to the 'sweet' thirds of mean-tone temperament, this pandering to a practice they had always disliked was something to be resisted as long as possible.

We may be sure, however, that the present supremacy of equal temperament will remain unassailable unless and until someone invents a really practicable means of playing, at will, one of two alternative notes on the organ by a single key (see Chapter 6, p. 54). Should that ever come about and bring admitted advantages in practice it will no longer be the primary aim of the organ-tuner to make twelve tempered fifths fit exactly into seven octaves. This brings us to the comma of Pythagoras, which is the excess of twelve true fifths over seven octaves.

The previous chapter explained, as a matter of arithmetic, how the comma of Pythagoras was eliminated in equal temperament. It did not, however, explain in detail just how the series of twelve true fifths departs from the notes of a truly-tuned twelve-note scale, and fails to end up on a high B♯ as our musical notation suggests it should do. That is now easy to explain, and we will set out the explanation because it gives a more understandable picture of the comma of Pythagoras than does the theorist's explanation that it is an interval with a ratio of $\frac{531441}{524288}$

As we have seen, to produce a truly tuned minor triad on the second-degree note of the scale of C we must use a D which is only a lesser whole tone above C. But the keyboard would provide no such note. There would be only one D in each octave; and it would be needed to produce a true fifth standing on G and would be a greater whole tone above C, as we found in the previous chapter. So on any twelve-note keyboard instrument whose notes are all tuned to give true intervals with a tonic C, the interval from D to A will be a comma less than a true fifth. To produce a true fifth we must add a comma (*k*) to the interval from D to A by tuning A a comma sharp, and the true fifth would then become D to (A + *k*).

This, however, is not the whole story. C♯ is a diatonic semitone below D and G♯ is a diatonic semitone below A. So we shall need C♯ to (G♯ + *k*) for a true fifth (G♯ + *k*), meaning G♯ tuned up a comma, and so on in what follows. Similarly, since E♭ is a diatonic semitone above D, and B♭ a diatonic semitone above A, we shall need E♭ to (B♭ + *k*) for a true fifth. We may now write out our twelve fifths with these modifications, it being assumed that, in such a series of rising fifths, octaves tuned downwards would be interpolated whenever necessary to enable us to remain in the middle of the keyboard.

1. C to G
2. G to D
3. D to (A + *k*)
4. (A + *k*) to (E + *k*)
5. (E + *k*) to (B + *k*)
6. (B + *k*) to (F♯ + *k*)
7. (F♯ + *k*) to (C♯ + *k*)
8. (C♯ + *k*) to (G♯ + 2*k*)

But by a chromatic change we can dispose of nearly all that 2*k*, for we learnt in the previous chapter that in a true tuning G♯ A♭ is a diesis, and we showed (in equation (6) therein) that two commas exceed a diesis by a schisma (s). So we may re-write the last fifth as:

8. (C♯ + k) to (A♭ + s)

and continue

9. (A♭ + s) to (E♭ + s)
10. (E♭ + s) to (B♭ + k + s)

But in the previous chapter we found that $c + s$ made up the comma of Pythagoras (p); so we may write the last fifth:

10. (E♭ + s) to (B♭ + p)

and conclude:

11. (B♭ + p) to (F + p)
12. (F + p) to (C + p)

The meanings of these fifths perhaps becomes clearer still when we remember that there are eleven schismas in a comma (k), so a schisma measures a slight mistuning. In fact, as a mistuned unison, it would produce only one beat in two seconds at the pitch of A♭ in the treble stave.

Examining our series of fifths we find that those numbered 2, 4, 6, 7, 9 and 11, have each leapt over a C, six C's in all, while the last one has reached the seventh C and gone past it by a comma of Pythagoras.

8

THE MYTH OF EQUAL TEMPERAMENT

No true harmonic ideas are based on equal temperament.

SIR DONALD TOVEY, *Encyclopaedia Britannica,* 14th ed. (article 'Harmony').

IT often happens that the piano is the first musical instrument with which a child becomes familiar. If it interests him, its seven white keys to the octave give him his first idea of the diatonic scale. If he learns to play a little on the piano, and to use some of the black keys for scales of different pitch, he is confirmed in his idea of a scale consisting of a fixed series of "notes"; and he thinks of the intervals between successive notes as being sometimes a semitone but more frequently a whole tone, all whole tones being, to him, the same size. As his experience grows and he learns to manage the key of A major or C minor, he thinks of each of the black keys as dividing the intervals between two white keys, musically as well as mechanically, into halves. He makes no distinction between G♯ and A♭, and he thinks of all semitones, diatonic or chromatic, as being exactly the same size and always half a whole tone. The notions implanted in childhood take deep root, and many of us never outgrow these early impressions; for a lifetime we continue to think of the tuning of the piano as a musical scale.

It is rather interesting to reflect that a child who was born a century or so before we were would probably have acquired a somewhat different conception. His parents' piano would have been tuned in mean-tone temperament, at any rate if he were an English child, and he would have found that A♭

in the key of E♭ major was not produced satisfactorily by playing G♯. Later, he would have been pleased to find that, on a few organs, the corresponding black key was divided, so that he could sound two separate sets of pipes; one for G♯, the other for A♭. He would certainly form an idea that diatonic and chromatic semitones are not the same size. He would think of the interval G♯—A (using different letters) as a diatonic semitone, and the interval A♭—A (using the same letters) as a chromatic one.

The musical amateur of to-day is often unaware that the tuning with which he is familiar, on his piano, became general in this country only a little more than a century ago, and is quite a modern innovation on the organ. The tuning of organs in mean-tone temperament was the rule in the first half of the nineteenth century, and our grandfathers contrasted it favourably with the new-fangled tuning. Thus, in *The Philosophy of Music* (1879), William Pole wrote:

> The modern practice of tuning all organs to equal temperament has been a fearful detriment to their quality of tone. Under the old tuning an organ made harmonious and attractive music, which it was a pleasure to listen to. . . . Now, the harsh thirds, applied to the whole instrument indiscriminately, give it a cacophonous and repulsive effect.

All musicians know that equal temperament is an acoustical compromise, tolerated by many ears on the piano, and designed to satisfy as completely as possible three incompatible requirements—true intonation, complete freedom of modulation and convenience in practical use in keyed instruments—and that it sacrifices the first of these to the second and third. But they are not always quite so clear about the precise effect which this compromise has on the various musical intervals. This effect is illustrated in Fig. 1. On the right-hand side the intervals of equal temperament, applied to the scale of C, are drawn with theoretical accuracy. On the left-hand side various musical intervals are drawn with a similar theoretical accuracy, an accuracy that ignores the limitations which different pairs of human ears experience in differing degree in their estimation of those intervals in varying circumstances that occur in musical performance.[1] The tempered fourths and fifths are nearly

[1] [Ref. 6], p. 445.

true, but the tempered thirds and sixths are faulty. The difference between an octave and a perfectly true major seventh, or between a perfect fourth and a major third, is a diatonic semitone; and it is obvious from the figure that the tempered semitone is appreciably less than a diatonic semitone. The difference between a major third and a major tone is a minor tone; and the figure makes it evident that the tempered whole tone approximates more closely to a major tone than to a minor tone. (This is where it differs from the mean-tone of the old tuning which was half-way between a major tone and a minor tone, with the result that the mean-tone semitone was larger than a diatonic semitone).

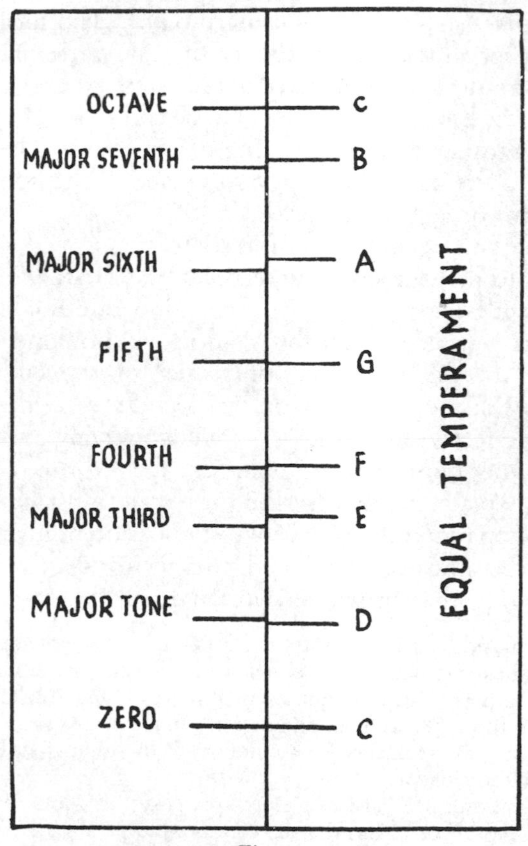

Fig. 1

Let us return, however, to our own childhood and think of the scale as we then imagined it. As we grow older a select few, gifted with great delicacy of ear, may learn to play a stringed instrument really well. Experience will teach them the truth, emphasised by Lionel Tertis in his *Beauty of Tone in String Playing*, that to play perfectly in tune constantly calls for the most intent listening. They will know that, in a violin player, only carelessness, or lack of skill, or inattention to his faculty of hearing will explain faulty intonation. When they are told that a musical scale-system must be a flexible thing if all concords are to sound in tune they will say: 'Perfectly true!' They will appreciate also the license enjoyed in the intonation of decorating notes which do not form an essential part of the prevailing harmony.[2] The idea that the estimation of musical intervals, by the ear, varies in accuracy with the musical occasion may or may not interest them, but they will not boggle at it. They will be conscious of something not quite satisfactory in the tuning of the piano. They will be sure that there is something wrong about the tuning of a harmonium or a chamber organ.

Others of us again, not so fortunately endowed by nature, may turn to the study of counterpoint. As we make ourselves familiar with the music of Palestrina and our English Tudor composers we find, as Stanford told us we should, that we are beginning to think in the pure scale[3] of sixteenth-century polyphony and to hear, with our mind's ear, ideal singers singing perfectly in tune. A scale consisting of intervals becomes the most natural thing in the world, and when Stanford explains to us why the pure scale must be a flexible thing if intonation is to be true we at once accept his statement. We shall be amused or irritated, in varying degree, by those who imagine that music has any use for the so-called 'just

[2] 'An animated intonation [on the violin] is no more mathematically true than an animated time-keeping is strictly according to the metronome'. Moritz Hauptmann, 'Letters', quoted in *Music & Letters*, Vol. XIX [Ref. 6], p. 443, October 1938. Hauptmann objected to the piano, as an instrument to accompany his violin, not because of its *faulty* intonation, but because of its *rigid* intonation.

[3] C. V. Stanford, *Musical Composition*, pp. 13–17. It is one of the minor offences of the 'theoretician' that he confuses the pure scale of sixteenth-century counterpoint with 'just temperament'.

intonation' which was confuted by Dr. Murray Barbour, a 'tuning' for which a more logical and descriptive name would be 'just temperament' (see Chapter 9, p. 89).

Others of us may happen to read, in books on sound as a branch of physics, of the vibrations of sounding bodies which we perceive as musical tones. We learn to calculate, as a matter of arithmetic, the ratios between two rates of vibration which produce familiar musical intervals. We accept the results of laboratory experiments, made perhaps with sirens, as fixing these ratios; and thus far, but no farther, we bring the ear into the picture. Now this is quite the proper approach to the problem of tuning keyboard instruments, for the tuner's procedure is to estimate the rate of beating which produces the appropriate degree of mistuning required for equal temperament. The problem of tuning is, essentially, one of physical acoustics. The knowledge we may thus have gained does not equip us, however, to deal with the misconceptions of 'theoreticians'[4] who proceed to make other calculations, about the vibrations corresponding to musical tones we hear, which leave out of account the ear's power of estimating musical intervals in varying musical circumstances. Assuming the 'theoretician' to know more science than we do, and that the printed word must surely express accepted authority, we are troubled when we are told that the music of the last two hundred years rests on a false foundation because (as the 'theoretician' alleges)[5] it was composed in equal temperament, or that equal temperament is implied in enharmonic change,[6] or that the twelve-note semitonal scale of modern composers can only be explained by equal temperament. We reach the conclusion that there is some sort of conflict between science and music, and we decide that here are unfathomable mysteries and that it will be best to stick to music.

The *Musical Times* for December 1939 contains a short article which is so apposite to our theme that it might have been specially written to illustrate it.[7] By the kindness of the

[4] See page 81 and footnote 16.

[5] [Ref. 6], p. 447.

[6] See quotation with footnote 8 on p. 48.

[7] *Musical Times*, No. 1162, p. 795, 'An Aspect of Bach', by H. Ian Parrott.

author, and the courtesy of the Editor of that journal, I am able to reproduce a part of what he says and the first two musical examples he quotes from Bach:

> The more one studies Bach, the more incredibly fine his 'linear' writing—to use a modern phrase—appears. Take every line separately and each seems absolutely right; put them together and they seem righter still. . . . How often is the full harmonic richness of the following [Fig. 2] appreciated? Bach enjoyed it so much that he repeated it three times in different keys:

BRANDENBURG CONCERTO NO. 1, SLOW MOVEMENT

Fig. 2

> And then there is this [Fig. 3] (a fine thing, spoilt in Mendelssohn's edition—perhaps not deliberately—by the substitution of C♮ for C♯):

'CHRISTE, DU LAMM GOTTES' ('ORGELBÜCHLEIN')

Fig. 3

> I feel that there is hardly a single composer since Bach who has not been foreshadowed in some way by him. And it is the number of comparatively *recent* composers which makes this fact the more remarkable.

The author advances a suggestion that is wholly convincing for, when pointed out, it leaps to the eye: that recent composers hark back to Bach because they write polyphony and are not afraid of dissonance. But the whole article invites the attention of readers of this journal; and because it is so musicianly the acoustical science implied in it is unimpeachable.

Let us examine the clash between B♮ and B♭ in the first example. Here we are surely harking back to music of an earlier period than Bach's; to music written before temperaments were invented. This is what the Editors of *Tudor Church Music* say about the scale-system of Taverner's day (first quarter of the sixteenth century):[8]

> If a modern musician is asked to think of the key of D, his mind will probably construct a framework consisting of two Ds, an octave apart, with a set of six notes between. In Tudor times this point of view had not been reached. Scales were then governed by the hexachordal system, which, in the key of D, would comprise the hexachord D, E, F, G, A, B♭ with an extension below to C♯ and B♮, and an extension above to C. From this conception arose the fact that, whilst to us there exists a choice (when the seventh degree of the scale is required) between C♯ and C♮, to composers of the sixteenth century the notes were not alternatives but equal and substantive members of the key, with quite different functions.

Because the scale discoverable in the polyphonic music of Taverner and his contemporaries has this hexachordal basis, English composers of his day found nothing unnatural in those harmonic clashes which are a characteristic feature of their cadences. The melodic line was everything: harmony was in the making, it was being formed by writing concurrent melody. Without implying that Bach knew the music of our Tudor composers, can any musician question whence came the scale-system discoverable in the quotation from the first Brandenburg Concerto (Fig. 2)? Bach summed up the achievements of all his predecessors, and his harmony derives from the contrapuntal procedures of the polyphonic period. Like his English predecessors, he took delight in this logical application of principles embodied in the hexachordal scale-system. He adopted a tuning for his clavichord which had first been invented more than a century before his day. But he did so because his genius impelled him to claim complete freedom of modulation. To suppose that by so doing he established a false foundation for all later occidental music is

[8] *Tudor Church Music*, Vol. I, p. xlii, from which this quotation is reproduced by kind permission of the Carnegie United Kingdom Trust.

to ignore all the lessons of musical history, and to imagine that scales came first and music afterwards. Had some supreme authority abolished all keyboard instruments from the earth in the sixteenth century and ever after, composers writing for voices, strings, or the orchestra, would have made music from which we could distil the same flexible scale-system, with its use of enharmony, as we discover in their compositions today. Bach imposed no tyranny on the intonation of all later music. On the contrary he re-asserted its right to freedom of intonation, and he tuned his clavichord to express his musical imagination as well as it could. Only the 'theoretician', thinking in terms of physical vibrations and a *rigid scale*, *i.e.* a 'tuning' not a musical scale, can reach any other conclusion.

Many will have heard the first-quoted bar of the Brandenburg Concerto (Fig. 2) if only on the wireless, without being uncomfortably conscious of the clash between B♭ and B♮. The contrapuntist, knowledgeable in the use of *musica ficta*, will think of the B♭ as being close to the A; of the interval in the sequence A—B♭—A as being much less than a diatonic semitone; and similarly for the sequence Ċ—B♮—C of the oboes. The string player will assume that the cellos would probably flatten the B♭; and, so far as the limited flexibility of intonation of his instrument permits, a skilled oboe player would tend to sharpen the B♮. Moreover the ear is assisted by the difference of timbre between the strings and the wind which helps it to hear the two melodic lines apart. Now try the passage on the piano: the effect in the first bar is hardly tolerable. In the middle of the keyboard B♭ has a powerful octave overtone, and the clash of physical vibrations will produce an unpleasantly harsh sensation in the ear. The 'note' of the piano is lacking in acoustical definition, as compared with that of the violin or of many combinations of organ stops. There is evidence that the pianist's educated aural perception, which is not that of a skilled violinist, leads him to hear the intonation of the simple concords on his instrument as something which he does not distinguish from the concords of the 'pure scale'. But the physical clash of the tempered B♭ and B♮, struck together, will defeat the tolerant aural perception of the most complacent ear. Bach was not writing for a keyboard instrument in this Brandenburg Concerto, and musicians will agree that the 'theoretician' is not entitled to assume that, in writing it, he was thinking in terms of the tuning of his clavichord.

Let us now look at the notes falling in the third crotchet beat of the first bar of Fig. 2. To our modern ears the oboes are still playing a chord of A minor, while the bass has already modulated into D minor. Does this foreshadow any *prima facie* ground for assuming that what the theorist conveniently calls *polytonal* music presupposes equal temperament? The explanation which Dr. A. F. Barnes—writing as a musician—gives, that music written in the modern twelve-note semitonal scale depends on the contrapuntal framework and the enharmonic latitude enjoyed, is surely more logical, and far more musicianly. It is infinitely better science.

Nor does the clash of C♮ and C♯ in the second example, Fig. 3, lead us to qualify the conclusions of the preceding paragraphs, notwithstanding the fact that the organ, with its sustained tones, may often give more acoustical definition to a note than does the piano. Observe that the C♮ is a sustained note. If we play this prelude on the organ we find satisfaction in the ruthless melodic line of the 'alto', in spite of the rigid intonation of the instrument, because the sustained C♮ falls into the background of our aural perceptions. Our ears are fully occupied in attending to the two lines of melodic motion.[9] Similar examples are found in other organ preludes of Bach. For instance in the setting of 'Nun komm, der Heiden Heiland' for manuals only (Fig. 4, where marked with an asterisk):

Fig. 4

Similarly in the organ prelude, 'O wie selig seid ihr doch, ihr Frommen' (Fig. 5), Brahms (1833–1897) writes:

[9] Observations which are pertinent are given in Appendix 2.

Fig. 5

These passages are written thus, not *because* of the rigid intonation of the organ (Bach's organ was tuned in a 'sharpened' form of mean-tone temperament, see p. 55), but *in spite* of it. What evidence is there in all this that composers write, as the 'theoretician' supposes, in equal temperament?

Modern science gives no support to the 'theoretician'. Laboratory investigations[10] show conclusively that, measured by the physical vibrations in the air, the melodic scale of the violin player (unaccompanied) is a flexible affair. In any contrapuntal writing the string player's feeling is for his linear melody; only in a concord on the accent of the bar will his sensitive ear impel him to produce a consonance with the other strings. Helmholtz and Lionel Tertis tell the same story (see Chapter 9, pp. 82–4). But this is not all. It is easy to play out of tune: it is a superhuman feat to play 'off the note' with exactly the mistuning required for equal temperament, for we may be sure that the player has no physical means of reproducing equal temperament with the accuracy with which a good string quartet can play in tune. Even the piano tuner with an ear trained to measure the desired dissonance, and with ample time to listen to the beats he produces, does not tune the instrument perfectly in equal temperament.

Fig. 6 is a trace of the actual rates of vibration one tuner produced in the piano, as recorded by the Chromatic Stroboscope. This is a scientific instrument which records, instantly and accurately, the rates of vibrations that travel through the air to our ear-drums from something that produces

[10] P. C. Greene, *J. Acoust. Soc. Amer.*, 1937.

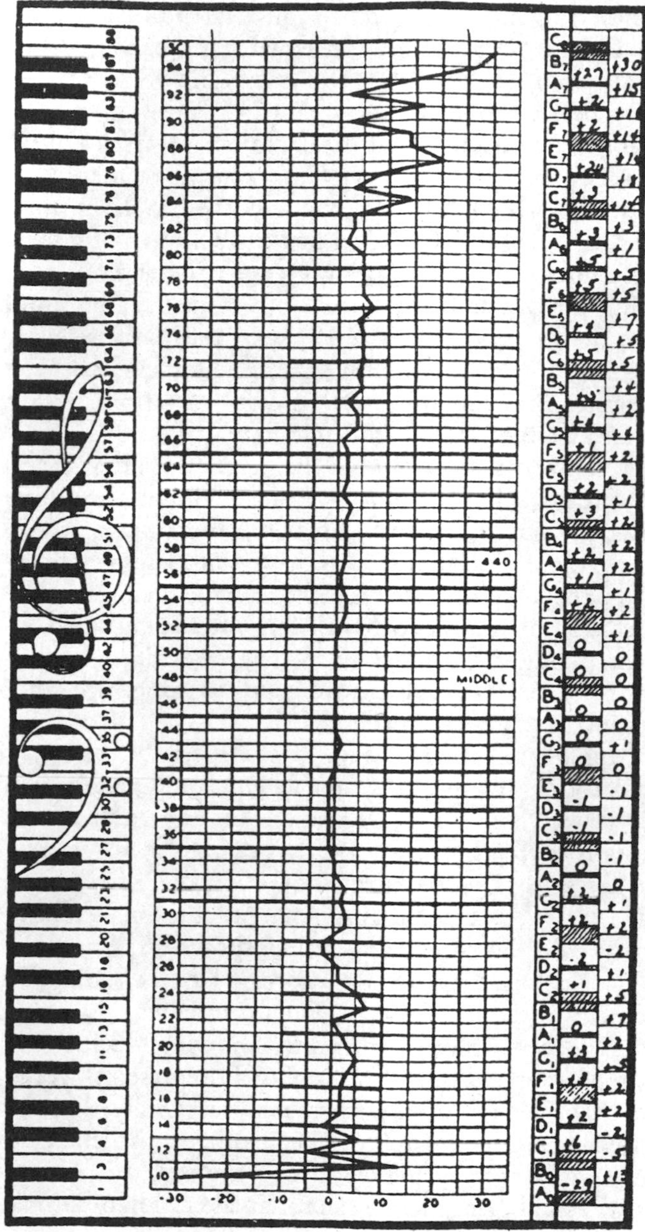

Fig. 6

a musical sound. The standard of reference is the nearest vibration of a perfect tuning in equal temperament of international pitch (A = 440). The rates of vibration which produce perfect equal temperament are represented by the vertical line down the centre of the figure. A vibration which is too sharp for perfect equal temperament is marked by a point to the right of that line: one which is too flat is similarly marked to its left. The broken line shown in the figure joins all these points in turn. The waggles in it therefore indicate changing deviations from equal temperament. The numbers marked at the foot of the figure give the measure of these deviations in *cents*, a cent being a small unit used to measure the relation between the rates of two vibrations. To give some idea of the size of this unit it may be added that the equally tempered fifth should be 2 cents flatter than the perfect fifth which is the objective of the violinist in tuning his instrument. The deviations shown are therefore of the same order as those deviations from true intonation which are implied in equal temperament. The actual deviations, in cents, for each key of the piano have been entered by hand in the columns on the right-hand side. The deviations in the top and bottom octaves, which are hard to tune, are of little significance for present purposes.

The tuning recorded in Fig. 6 was specially selected as a particularly close approximation to equal temperament by a skilled tuner; and for the selection of this record and for permission to reproduce it, the writer desires to express his indebtedness to Dr. Robert Young, of the firm of C. G. Conn Ltd., Elkhart, Indiana, U.S.A., the makers of the Chromatic Stroboscope. This record shows surprisingly good tuning, of a Steinway piano: records of other tunings show substantially greater deviation and a more waggly line, one of the less accurate ones constantly showing fluctuations by 5 cents or more, though the tuning it represents would be accepted as adequate. Not all tuners are equally skilled, and some pianos are harder to tune than others. An enharmonic change, which is very perceptible to the sensitive ear of the violinist, would theoretically call for an alteration of intonation by some 20 cents.

When the 'theoretician' assures us that composers or artists think or play, in any exact sense, in equal temperament we are entitled to ask him to explain, scientifically, how they do it. If the piano tuner does not succeed in attaining a perfect equal temperament, how can the artist, playing a momentary note on an instrument of free intonation estimate *exactly* a dissonance which, as Helmholtz showed, must lack definition? The 'theoretician' may be referred to Hauptmann's comments:[11]

> Do you know that Spohr maintains that the singer should learn intonation from a piano in equal temperament?— !— ?— :—; what marks of admiration shall I use? The fitting exclamations have yet to be invented. . . . And why should the singer cultivate temperament? . . . Thanks to the indestructibility of natural organisation *it cannot be learnt.* [The italics are the present writer's.]

And if the 'theoretician' replies 'well, of course, I don't mean *exactly*', has he not conceded his whole case? For that admits a flexibility of intonation in the musical scale of an instrument with free intonation. The artist whose sensitive ear becomes conscious of the dissonance of a mistuned interval will use enharmonic change wherever it is needed to enable him to hear his intervals 'in tune'.

The author of an article on 'The Schoenberg Concept'[12] in *Music and Letters* for April 1939 was troubled by the difficulty of reconciling two truths alleged by the 'theoretician', 'the chord of nature' and the handicap of equal temperament under which all occidental music has laboured since Bach's time. Those of us who have encountered this difficulty, and found that both truths are imaginary, will sympathise with him; and any resentment we may have against 'theoreticians' for misleading honest musicians will be intensified. The outstanding feature of temperament is, not the precise degree of mistuning implied in any particular system, but the fact that it postulates a *rigid* intonation. The idea that Bach, of all people, elected to compose in such an intonation has only to be put into words to be dismissed as untenable. Mr. Noel Heath Taylor tells us that, in reply to a question, Schoenberg said, with a shrug: 'Twelve-tone equal temperament is

[11] *Letters of a Leipzig Cantor*, Eng. trans. (Novello), Vol. I, p. 150.

[12] *Music & Letters*, Vol. XX, No. 2, p. 184, *The Schoenberg Concept*, by Noel Heath Taylor.

practical. There is no other popular medium available to the composer today'. No one will quarrel with the natural meaning of that statement; but the answer we want is that to the next question: 'Do you consider that modern composers think, and that their music should be played, with a rigidly fixed intonation?' There is a quotation at the end of the article which suggests one reply that Schoenberg might make: 'The criterion for the acceptance or rejection of dissonances is not that of their beauty, but rather only their perceptibility'. Exactly! What matters is the intonation we hear, the product of aural perception, not the vibrations we listen to. For though it is the latter which determine a physical tuning, they represent only an approximation—which may or may not be infinitesimally close—to the intonation we hear. Here, at any rate, Schoenberg's theory and practice find a common factor in scientific truth.

On the other hand the 'chord of nature' is a scientific misconception which is at least a hundred years behind the times. Experiment taught the Greeks that consonances were produced by vibrating strings whose lengths had simple ratios — $\frac{1}{2}$ for the octave and $\frac{2}{3}$ for the fifth. For over 2,000 years this relation between consonance and whole number was a puzzle. Before Rameau (1683–1764) no one found the right approach to the only explanation we can offer today. It was Rameau who observed that the first, third and fifth of the harmonic vibrations of a string, with suitable octave transpositions, gave all the 'inversions' of the major triad. Rameau's observation, and the theory about *générateurs* which he based on it, were set out lucidly and logically by a distinguished physicist and mathematician, d'Alembert (1717–1783). At the end of Part II of *Sensations of Tone* Helmholtz gives an interesting review of Rameau's attempt to find a *natural* explanation of harmony. He writes:

> No one who considers the great perfection and suitability of all organic arrangements in the human body would deny that when the existence of such natural relations have been proved as Rameau discovered between the tones of the major triad they ought to be most carefully considered, *at least as the starting points for further research*. [The italics are the present writer's.]

. . . .

> This attempt of Rameau and d'Alembert is historically of great importance, in so far as the theory of consonance was thus for the first time shifted from metaphysical to physical ground. It is astonishing what these two thinkers effected with the scanty materials at their command. . . . If I have been able to present something more complete, I owe it merely to the circumstance that I had at command a large mass of preliminary physical results, which had accumulated in the century which has since elapsed.

This is too modest a claim for the revolution Helmholtz effected in musical theory. He, in turn, shifted it from a physical to a physiological basis, by explaining how the overtones of a musical instrument enable our ears to distinguish between consonance and dissonance;[13] in short, how we tell whether two notes are in tune and why intervals differ in definition. This is a far cry from Rameau; and the passage from *Sensations of Tone* quoted above, also records facts, evident to d'Alembert more than 150 years ago, to which the 'theoretician' closes his eyes:

> No one knew better than d'Alembert himself the missing links of this system. Hence in the preface to his book he especially guards himself against the expression: 'Demonstration of the Principle of Harmony', which Rameau had used. He declares that so far as he himself is concerned, he meant only to give a well-connected and consistent account of all the laws of the theory of harmony, by deriving them from a single fundamental fact, the existence of upper partial tones and harmonics, *which he assumes* [the italics are the present writer's] without further enquiry respecting its source. He consequently limits himself to proving the *naturalness* of the major and minor triads. He does not mention beats, the real source of distinction between consonance and dissonance.

Others, however, lacked d'Alembert's accuracy of thought. As Sir Donald Tovey observes:[14]

> In England Rameau's doctrine raged unchecked by taste and common sense, and culminated in Dr. Day's famous application of homœopathy to the art of music.

The 'theoretician' who still derives his scale and his harmony *directly* from the harmonic overtones of musical instruments

[13] Not to be confused with concord and discord (wherever the musical idea of the time may draw the separating line). The 6–3 concord D—F—B is an acoustical dissonance, while the major 6–4, which was a discord in the polyphonic period and the next two centuries, is acoustically the most consonant triad.

[14] *Encyclopaedia Britannica*, 14th ed. (article 'Harmony').

neglects all that science has discovered about 'hearing' since Rameau's day. The very words 'chord of nature' are misleading. The harmonic overtones of a musical instrument are pure tones. If a group of them, produced electrically in the laboratory, are sounded together we do not hear a chord. Harvey Fletcher has shown[15] that if four consecutive upper partial tones are sounded together what we hear is a musical tone of the pitch of the fundamental, though that is lacking in the physical vibrations in the air. Ellis's 'duodenation' would lead musical composition to retrace its steps to Handel's day. Joseph Yasser's speculations in 'A Theory of Evolving Tonality' have at least the merit that they are informed by a true musical conception, that of a living and constantly changing art: he produces a most elaborate edifice, but its foundations rest in quicksand, for he derives the hexad of his supra-diatonic scale, its 'common chord', from the eighth, ninth, tenth, eleventh, thirteenth and fourteenth partial tones of a musical instrument. It would be instructive if we could persuade some physicist to sound pure tones of these pitches, simultaneously, and observe what is heard. In fact, however, only the merest traces, if any, of the vibrations corresponding to these high partial tones are to be detected, by laboratory analysis, in many of the complex physical vibrations in the air which produce the tones we perceive as the sounds of musical instruments. The normal human ear cannot detect them. If future composers should ever develop the scale in the direction Yasser indicates, its intervals will depend, as Schoenberg would say and as he himself recognises, on their perceptibility, and not *directly* on a series of high harmonic overtones such as are often lacking from a musical sound.

The use of the word 'chord' suggests to the musician a group of notes played on musical instruments. But that is something quite different from the supposed 'chord of nature', for each of the 'notes' of musical instruments contains its own overtones which give it definition and therefore individuality. Were there any truth in the conception of the 'chord of nature', using the words in their plain meaning, composers would long ago have given up writing for anything so harsh and crude as the orchestra would then be. Their genius would

[15] *J. Acoust. Soc. Amer.*, 1934, **6**, 67.

have impelled them to explore, instead, the nebulous beauties of a consort of ocarinas, for the note of the ocarina is a practically pure tone.

The 'theoretician' is not entitled to stop just where he chooses in the application of his theory. All its logical implications must be explored. Should that exploration prove too much his theory stands self-condemned, as Helmholtz pointed out. The fever of Rameau's doctrine still rages; all we can do is to prevent the infection from spreading. Helmholtz told us that 'we must distinguish carefully between composers and theoreticians [*musikalischen Theoretikern*]'.[16] Equally necessary is it to distinguish between men of science and 'theoreticians'. The notion that harmonic tissue is derived *directly* from the harmonic overtones of a musical instrument, and its corollaries, a physical basis for music and a scale with fixed intonation, are the peculiar characteristics of the 'theoretician'. We may use them to identify him (and his victims), just as we tell the leopard by his spots.

[16] *Sensations of Tone*, Eng. trans. of 1875, p. 345, and [ref. 6, p. 446]. 'Theoreticians' neglected the evidence of the composer's ears.

9

JUST TEMPERAMENT

THE year 1938 saw the publication of a slender volume which makes an authoritative contribution to our knowledge of musical acoustics. It has particular interest for those who are perplexed by a conflict of view between musicians and theoreticians concerning intonation, the musical scale and temperaments. For them it is the more instructive because the words 'acoustics' and 'scale' are not to be found in it. In the author's words this book is 'exclusively devoted to a proved method for acquiring expressive beauty of tone'. It is written by Lionel Tertis, and its title is *Beauty of Tone in String Playing*. Here are quotations from the first page of text:

> Perfect intonation is the rock-foundation of the string player's equipment. Without this, no one should be allowed to perform in public.... A 'good ear' can become permanently perverted by negligent, superficial, non-penetrative listening on the part of the performer. This inattention to one's faculty of hearing is a vice of such rapid growth that in a very short time the player admits faulty intonation with equanimity, becoming quite unconscious that he is playing out of tune. The certain road to never-failing perfect intonation is listening of the most *concentrated* kind.... The moment a note sounds in the slightest degree out of tune, *correct it immediately*. You will notice how much richer is the sound of a note that is absolutely in tune.

The student of Helmholtz who reads these words will turn with particular satisfaction to the pages of *The Sensations of Tone* in which that great scientist discusses quartet playing on stringed instruments. The following passage occurs:

> When quartets are played by finely cultivated artists it is impossible to detect any false consonances. To my mind the only assignable

reason for these results is that practised violinists, with a delicate sense of harmony, know how to stop the tones they want to hear, and hence do not submit to the rules of an imperfect school. That performers of the highest rank do really play by natural intervals, has been directly proved by the very interesting and exact results of Delezenne.

By 'an imperfect school' he referred, as an earlier passage shows, to a school of violin playing which, since the time of Spohr, had aimed at producing (as was supposed) equally tempered intonation. The context—like the rest of the more musical part of his book—shows that, by 'just intonation' he intended the musician's meaning, 'playing in tune', which applies primarily to consonances. He certainly did not intend us to think of the theoretician's mistaken conception of just intonation as a method of tuning the white keys of a keyboard instrument to give correct major triads on the tonic, dominant and subdominant; the conception which Dr. Murray Barbour 'confuted' in an article he contributed to the January 1938 issue of *Music & Letters*, when he met the theoretician on his own ground. Helmholtz's 'justly-intoned' harmonium was not limited to such a tuning. It had two manuals which were tuned to produce, between them, fifteen major triads and as many minor ones with perfect thirds and with fifths which, in theory, had one-eighth of the error of the fifth of equal temperament,[1] but, in practice, were tuned true, because this small error was not discernible by the tuner's ear. It gave considerable, though not complete, freedom of modulation.

Observe that, in the passages quoted above, both authorities lay emphasis on playing in tune as an outstanding mark of good string-players. Even more significant is the emphasis both lay on the power of a 'good ear' to appreciate pure consonances. Our theoretician made the fatal error of leaving the ear out of the picture altogether, and the tuning which he miscalled just intonation was based, as Dr. Murray Barbour explained, on an arithmetical conception of the scale. This conception began and ended with the rates of vibration of musical sounds and ignored the ear's power of apprehending them.

In an article on intonation in *Music & Letters* for October 1938 [Ref. 6], the present writer attempted to illustrate the

[1] The *Sensations of Tone*, Eng. trans. of 1875, p. 492.

varying sensitiveness of the musical ear for intervals, melodic or harmonic, in varying circumstances that occur in the performance of a composition. This varying sensitiveness of the musical ear was expounded by Helmholtz; and his exposition makes it evident that the ear is satisfied with what are only approximations—more or less close—to the just intonation of ideal intervals. Not only, however, does the response of a particular pair of ears vary with different musical circumstances but, as Lionel Tertis tells us, there are natural differences in the power of different pairs of ears to detect faulty intonation. Not all of us have the 'good ears' of the string player. We can educate such natural powers as we possess, or we can ruin them by carelessness, but we cannot create them.

Dr. Murray Barbour showed that the method of tuning he described breaks down at once when it attempts to produce a concordant triad on the second degree note of the scale. This is decisive. For example, it would make the second concord of 'Sumer is icumen in' sound out of tune (see also Chapter 5, p. 41). The arithmetical calculations which are generally employed to expound this method of tuning are often too unreal to the musician to carry much force: he may therefore fail to appreciate the significance of this observation. A diagram that shows musical intervals will be more convincing. Such a diagram is given in the Figure, in which two sets of graduations, one left-handed the other right-handed, are fitted together. On each set the different musical intervals, all measured from the zero, are drawn in their theoretically correct proportions. On the left-hand side the major sixth and fourth are marked, in brackets, as A and F of the scale of C major. On the right-hand side the graduations for the fifth and the minor third are adjusted to these notes. The zero does not come opposite D on the left-hand side. For a perfect consonance D must be moved down till it is only a minor tone above C. It becomes a mutable note. As Helmholtz pointed out, not only D but B♭ becomes a mutable note in the scale of C. For the relative minor, D and G have to move down by the difference between a major and a minor tone and B and G become mutable notes. We are, of course, thinking only of *intervals* that are tuned with ideal exactness. The notes marked on the left-hand side in brackets give the tuning which

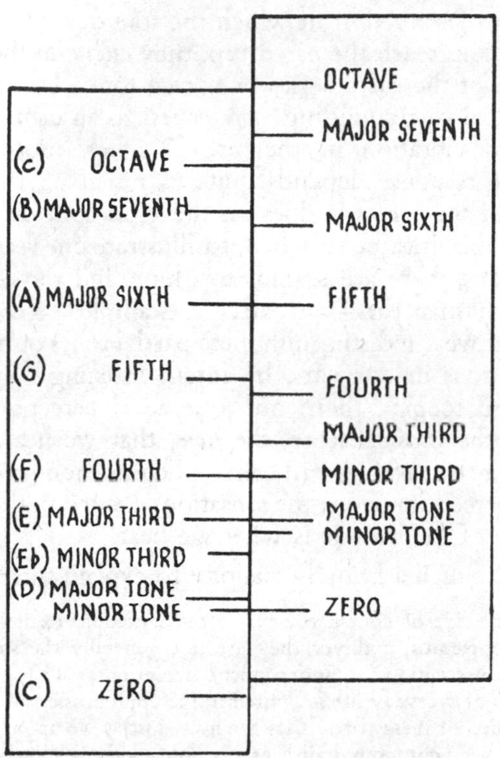

the theoretician called 'just intonation', which is evidently limited, for practical use, to playing Amens or their decorated equivalents. Like Dr. Day's theory of harmony, this effort of the theoretician to construct a musical scale from scientific data is to be treated, today, as a museum specimen. Unfortunately, the respect with which it was accepted in the text-books of a past generation, written by those who imagined that scales came first and music afterwards, tends to hide the fact that it never was a musical scale. A scale as evolved by the simplest counterpoint must be flexible. The pure scale of strict counterpoint must contain mutable notes.

Consider what is implied in the words used by Lionel Tertis, 'the sound of a note that is absolutely in tune'. The note is sounded by his viola. Travelling through the air to our ears it takes the form of minute periodic vibrations, of a complicated character, in the air itself: to-and-fro movements along a line pointing in the direction in which the sound is going. When these vibrations reach the ear-drum, they excite in the nervous mechanism of the ear a series of partial tones, each of which is a pure tone corresponding *more or less* to an element of the complicated vibration in the air. We say 'more or less', because the response depends quite as much on the sensory apparatus of the ear as it does on the vibration. It is easy to realise that this must be so when, to illustrate one reason for it, we recall that there are sounds too high and too low to be audible to human ears—an extreme example, of course. To some extent we can distinguish these partial tones or overtones, these sensations in our ears, by intent listening. But, unless we try hard to pick them out, our aural perception blends them into the musical note, the *tone*, that we hear. A tone then is what the psychologist calls a phenomenon produced by our aural perception from the sensations excited in the nervous mechanism of the ear. It is what we hear.

Helmholtz used a homely analogy to explain this:

> Partial tones are of course present in the sensations excited in our auditory apparatus, and yet they are not generally the subject of conscious perception as independent sensations. The conscious perception of everyday life is limited to the apprehension of the tone compounded of these partials, as a whole, just as we apprehend the taste of a very compound dish as a whole, without clearly feeling how much of it is due to the salt, or the pepper, or other spices and condiments.

And just as we ourselves can add salt, pepper and other condiments to suit our taste, when the dish is served, so the ear can, and does, add partial tones and combination tones, as a result of its own unsymmetrical construction; and it hears a tone which does not correspond exactly to the vibrations in the air. It can go one better: it can remove some of the condiments, and even part of the meat put in by the cook; for the ear is less sensitive to the deeper bass notes than it is to treble ones. The musical note we perceive, the tone, though related to the periodic vibration in the air, is quite different from it in both nature and degree. To liken the difference to that between chalk and cheese would do it an injustice. Rather should we say that the difference is that between chalk and the taste of the cheese.

Does the same cheese taste the same to different people? Do differences in our taste buds, or in our perception of their sensations, affect the taste of the same cheese to different people, much as different makes of cheese affect the taste of any one of us? Judging from our other senses the answer is almost certainly 'yes'. Here again, as we have observed, Lionel Tertis enforces a point about aural perception which is analogous. The powers of different ears, whether as the result of continued carelessness, or of the limits of natural gift, affect the perception of intonation by their owners. We are reminded at once of the question: Does the same music sound the same to different people? We are beginning to realise that it should be stated: How far does the same music sound different to different people?

Modern scientific research throws yet other side-lights on this bewildering matter of aural perception; for example, the creation, by the ear, of a 'fundamental' from selected 'overtones' when no vibration corresponding to the fundamental exists in the air. Organists will recall the 'acoustic bass' sometimes met with on the pedal organ. Difference tones may help to create this fundamental, as may be true of a church bell; but the loudness of the fundamental, like that of the strike-note of the bell, suggests that it is a product of aural perception. Sound-curves have been published showing the vibrations produced in the air by the open G string of a violin in which the octave vibration has about 100 times the intensity of the prime; but the ear would produce a proper G tone from the overtone structure.

This is one illustration of what we are learning about aural perception; there are others equally well established. We naturally think of the pitch of a musical sound that we hear as something that is fixed by the rate of the vibration in the air. But men of science tell us that this is not the whole story. If a pure tone is produced, softly, which the ear hears as being of the pitch of A♭ at the top of the stave with a bass clef, and if it is then sounded very loudly without altering the rate of vibration, the ear will say that the pitch has fallen. Some ears will discover a fall of pitch by as much as a minor third. This is almost unbelievable, but it is unquestionably true. If we add harmonics, however, the fall of pitch is greatly reduced, perhaps to one-fifth of what it was with a pure tone. Pitch then is a perceptual effect. It depends, to some extent, on the intensity of the vibration as well as on its rate, and it also depends on overtone structure. Pitch must be distinguished from frequency, just as loudness must be distinguished from intensity. Frequency and intensity are measures of the vibration in the air. Pitch and, still more, loudness depend on what the ear makes of the vibration.

Those theoreticians who dealt only in vibrations left the ear out of the picture in a very real sense. What the man of science calls the correlation between the complex vibration in the air and the tone heard is by no means simple. It depends on the nature of the response of the auditory apparatus of the ear, and on aural perception. We cannot, like the theoretician of fifty or more years ago, take it for granted. We must find out the facts. We know far less about them than many people imagine; and not only does the musical ear hear (*i.e.* perceive) musical sounds which the vibrations would regard as caricatures of themselves, but the caricature varies in detail from individual to individual. If we assume that this never has any effect on our aural perception of intonation we are just guessing, and on occasion we are demonstrably wrong.

The theoretician's mistaken conception of 'just intonation' as a tuning that produced a scale was faulty both as music and as science. What then are we to do about it? The theoretician turned a deaf ear to music, which makes it tempting to poke fun at him. A more constructive attitude, however, is suggested by an entry in the first edition of a glossary of acoustical terms prepared by the British Standards Institution (B.S. 661 : 1955). The musical part of this glossary is very useful as an attempt to explain, to the musician on the one hand and the scientist on the other, the different languages each uses. Such a glossary cannot initiate terminology. It can only record terms which are in established use and co-ordinate their meanings so as to avoid, as far as possible, confusion and duplication. It is therefore interesting to find that there is an item in the glossary: 'Major scale of just temperament' and underneath it, in smaller type, 'Major scale of just intonation'. The scale is defined as having '8 tones to the octave, with frequencies consecutively proportional to 24, 27, 30, 32, 36,

40, 45, 48', exactly the tuning postulated by Dr. Murray Barbour[2] Some musicians may be disposed to question the propriety of following the practice of those who have called this just temperament. But, in fact, were they not right? A temperament is a method of tuning a keyed instrument, in particular a keyboard instrument with fixed intonation. Its essential feature is this rigidity of intonation, which is precisely why it is not a musical scale. As a result of this rigidity, certain intervals are mistuned in a manner which aims at a musical objective. In the modern tuning of keyboard instruments, called equal temperament, every interval except the octave is mistuned to attain a musical objective: complete freedom of modulation. An earlier temperament gave results that were in some ways more satisfactory, but it had a more limited objective. It was evolved some six hundred years ago, and in this country it remained in use by some tuners until well into the nineteenth century. It was satisfactory so long as modulation was limited to a few keys; for though the fifths were flatter and the fourths sharper than those of equal temperament, it produced perfectly tuned major thirds. It is called mean-tone tuning.

'Just temperament' now falls into its logical place in this scheme. It differs in detail, but not essentially in nature, from the other temperaments. It produces perfectly tuned intervals for the major triads on the tonic, dominant and subdominant. There its practical usefulness ends. So limited is its musical objective that it cannot be used for real music, as the Tudor musician well knew. Its true objective is to serve as a theoretical tuning, to be used as an ideal standard with which to compare, in terms of vibrations, and vibrations alone, the intervals of mean-tone tuning and equal temperament. It is a measuring rod and nothing more. Were all musicians to agree, forthwith, to call it 'just temperament' they could expect to find in the next edition of the glossary that 'just temperament' was a tuning (not a scale), and that 'just intonation' meant playing or singing in tune. The musician would then come into his own. Music, to him, is what he hears (or perceives), not a matter of vibrations. When he wants to know whether string players are playing in tune he accepts as final the decision of a

[2] *Music & Letters*, XIX, p. 48, January 1938.

great artist. If a Lionel Tertis tells him that they are playing in tune, they *are* playing in tune. Whether they conform to some physical test of the theoretician does not matter in the least. All that matters is the judgment of the trained musical ear.

The general adoption of a nomenclature which was musically consistent would be a step towards the removal of a difficulty encountered today by the student of music and the scientist alike. Scales are made by composers, not by theory. To find out what the scale system really is we must turn to musical composition. The musical scale of the sixteenth century was a flexible thing. Its flexibility was increased when composers of the classical period made use of enharmonic change—for, as Sir Donald Tovey assures us, enharmonic modulation presupposes just intonation. The flexibility of the musical scale has been further increased in the twentieth century by the enharmonic latitude of the twelve-note semitonal scale. In short, a scale is merely a method of classifying and labelling the musical material used by composers and skilled artists. It can never be produced by a keyboard instrument unless the aural perception of the trained musical ear enables it to hear the tuning as something different from its physical original. The available evidence suggests that many of us hear the intonation of the piano in this way.

Discussion of nomenclature prompts one further comment. The word 'tone' is used in musical acoustics in five different senses, and each has appeared in the preceding paragraphs with the support of competent authority! The musician is responsible for two. First, it means a musical interval. Second, it refers to the quality of the musical sound that is heard, as in 'Beauty of Tone in String Playing'. But the scientist has added three other meanings. How much easier would musicians find it to follow these meanings had he adopted three different spellings! For example, he might have written about partial *toans*. The term as used for the musical sound we hear, corresponding to the complex dish with its condiments described by Helmholtz, he might have spelt *ptone*. And when he wanted to change the meaning altogether and refer to vibrations in the air, he might have adopted a Spanish device and written of *toñe*. No one will dream of adopting these obviously convenient spellings; but before the reader dismisses them as altogether too flippant, let him consider whether they might not have a mnemonic use which would help him to remember what we have been discussing. The *a* in to*a*n, is the initial letter of *a*uditory. It reminds us that a partial tone belongs essentially to pure auditory sensation. The *p* in *p*tone, the musical sound we hear, reminds us that a tone is what the *p*sychologist calls a *p*henomenon *p*roduced from these sensations by aural *p*erception. Finally the spelling to*ñ*e, reminds us that we are dealing with a use which is foreign to the word tone, and that we are now thinking of what Helmholtz described as a sum of pendular vibrations in the air, each with its own characteristic frequency; for science uses the sign ~ as shorthand for cycles per second, that is frequency, or rate of vibration.

Is all this of any importance for musicians? For the composer and the skilled artist, it has no importance. Their instinct is a true guide. Those who are at all interested in theory, or in getting rid of misconceptions for which we have to thank a past generation of theoreticians, will be able to judge of its significance for themselves. But there is another reason why the problems of musical theory should make an appeal to some of us. Science is at last beginning to resolve the difficulties which surround our conceptions of the musical sounds we perceive. The man of science, working at problems of aural perception, needs the co-operation of the musician. In the history and practice of music there is to be found much evidence which the man of science needs to use in his investigations; for if we are to understand musical acoustics we must begin with music and establish correct musical premises.

10

THE LESSON OF MEAN-TONE TUNING

THOSE who consult the article on Temperament in the fourth edition of *Grove* find that this is 'the name given to various methods of TUNING'. [The corresponding article in the fifth edition is by Ll.S.L.] They go on to read descriptions of tunings in equal temperament and mean-tone temperament, and find that the article then comes to a somewhat abrupt stop with the following editorial note:

> In former editions of this Dictionary the writer of the above here developed at length an argument for the partial restoration of mean-tone temperament. This argument was based largely on the researches of Bosanquet,[1] and included a description of the generalised keyboard [he] invented.... Fifty years of experience since the article was written, however, show that music has not taken the directions indicated.

Here we find the appropriate epitaph for the conscientious efforts made by nineteenth-century theorists to counter the supposed injuries which equal temperament was going to do to the art of music. The complicated keyboards of Peronnet Thompson (1783–1869), Colin Brown, and others, and the ingenious one designed by Bosanquet, referred to in the original article, become museum specimens. That is not to

[1] For the tuning of his experimental organ Bosanquet used Nicholas Mercator's temperament, which divided the octave into 53 equal parts. But this division of the octave is of much earlier date than Mercator's day (*c.* 1630–1687). Yasser states in his *Theory of Evolving Tonality*, p. 31, that a Chinese scholar, King-Fan (*c.* 200 B.C.) 'constructed a system of 53 tones within the compass of an octave'.

accept Ellis'[2] view that the great composers from Palestrina to Wagner thought in tempered music. Lecky, who wrote the article in question, had a much clearer and more musicianly conception. His insistence that temperament is a 'method of tuning' has increased significance when contrasted with the *a priori* assumption of other writers of the period, that the tuning of the ubiquitous piano necessarily identified the musical scale with equal temperament. Today, when we have at our disposal exact scientific measurements which show that, as judged by the physical vibration in the air, the piano is never tuned in a theoretically accurate equal temperament, that its intonation shows frequent deviations from the theoretical ideal, that these deviations though normally small are not the same for different pianos and different tuners, that many of the fifths are true while many of the octaves are not, and that there is constant stretching of the octaves particularly in the extreme bass and treble, we naturally hesitate to accept the theorist's account of the damage done to the art of music by the system of tuning the piano. We are led to consider, instead, the problems of aural perception presented by the piano as heard by different musical ears, and the possible discrepancies between the vibrations in the air and the trained musician's perception of the response they excite in his ear. We brush aside all the pseudo-scientific speculation of the *a priori* theorist, and we begin afresh by first considering music as produced away from the keyboard and, in particular, the intonation of the string quartet and that of unaccompanied voices singing sixteenth-century music. We find that, from the beginning of the sixteenth to the end of the nineteenth century, the musical scale as judged in this way was always a flexible affair. Temperament, as a tuning, was an attempt to approximate as nearly as may be on an instrument of fixed intonation to the flexible intonation of strings or unaccompanied voices.

The original article in the first edition of *Grove* is a most musicianly piece of writing; but it exhibits one grave defect of the musical scholarship of the day. The writings of the nineteenth-century theorists are almost invariably found

[2] Alexander Ellis (1814–1890), F.R.S., translated Helmholtz's *Tonempfindungen* into English as *Sensations of Tone*, but unfortunately added explanatory footnotes of his own.

wanting when tested by the music of the sixteenth century, which was a closed book to most of them. Examples will readily occur to readers of the original article in *Grove*. Perhaps the fundamental error of these theorists was to assume, without qualification, that a note, once sounded, must maintain its intonation unaltered so long as it lasts. That assumption fails when tested by the music of Palestrina, a theme which I have developed at length elsewhere.[3] Among modern writers, Stanford was insistent on the existence of mutable notes in the 'pure scale' he distilled from sixteenth-century polyphony. This term means something more than the existence of two forms of the same note to be used as alternatives when the music so requires. The note *as heard* in the singing of a good choir must really be mutable. And this reminds us of Tovey's observation[4] that on the piano we imagine a change of intonation in an enharmonic modulation, an observation which points to an intriguing problem in aural perception.[5]

Those who are interested in the flexible intonation of strings will find much food for thought in the article 'Temperament' in Cobbett's *Cyclopedic Survey of Chamber Music*, which discusses when and how the string player tempers his intervals. But earlier writers were equally clear about the intonation of strings and unaccompanied voices. Robert Smith, writing his *Harmonics* in the middle of the eighteenth century, quotes with approval a remark of Huygens (1629–1695):

> Mr. Huygens observed long ago, that no voice or perfect instrument can always proceed by perfect intervals, without erring from the pitch at first assumed. But as this would offend the ear of the musician, he naturally avoids it by his memory of the pitch, and by tempering the intervals of the intermediate sounds, so as to return to it again.[6]

Every student of sixteenth-century music learns that Palestrina's technique was skilfully devised to enable singers to

[3] [Ref. 26].

[4] Recorded in the article 'Harmony', *Enc. Brit.*, 14th ed.

[5] This problem was more fully indicated, by the present writer, in the final essay in *The Musical Ear* (Oxford University Press) [Ref. 11].

[6] Robert Smith's discerning comments were set out, by the present writer, in *Phil. Mag.*, Ser. 7, Vol. XXXIV, p. 476 (July 1943) [Ref. 27]. Dr. Robert Smith, F.R.S., was Master of Trinity College, Cambridge, and founder of the Smith's Prizes.

sing beautifully in tune without loss of pitch. Fig. 1 provides a simple illustration. For convenience of reference I have barred this in $\frac{2}{2}$ time. The Mass is written in the Ionian mode transposed and, again for convenience of reference only, we may think of this phrase as in the key of F major. The intonation is governed by the ascending scale in the bass and tenor. In the second bar we meet the minor triad on the second degree of the scale, which requires a mutable note as we shall learn later. The result will here be found in the alto. The theorist may speculate on the way in which the singer will temper his intervals to sing in tune and maintain pitch, on whether the major third marked by a square bracket at *,

Palestrina (Missa Brevis)

Fig. 1

which is an 'intermediate interval' not part of the prevailing harmony, would be given melodic intonation as a Pythagorean third. The contrapuntist would point to the way in which Palestrina approaches and quits the leap of a fourth as the significant factor in maintaining pitch. The experienced choirmaster whose choir has been trained to sing unaccompanied would be quite sure that they could sing this phrase perfectly in tune, without loss of pitch. The significance of Huygens' acute observation becomes clear. But Ellis, whose complacent belief in his own theories was made possible only by his ignorance of musical technique, brushed aside this observation on the ground that mere memory was not enough.[7] Of course it is not enough. Memory is no sufficient protection against loss of pitch if the composer does not know his job. Composers who have trained themselves by the discipline of the contrapuntal technique of the sixteenth-century masters can write music, however modern, which is grateful to sing in tune. There is a most instructive piece of self-criticism, bearing on this very point, in Hauptmann's *Letters*

[7] *Sensations of Tone*, 1st ed., Translator's Appendix, p. 789.

of a Leipzic Cantor (Eng. trans. published by Novello). Finding that his choir lost true intonation in a passage in his *Salve Regina* he wrote:

> There is no justification for a composer who makes a pianoforte accompaniment indispensable for the performance of choral music; and the old masters were far from wrong, in adhering to a very peremptory code of laws, to regulate such compositions as this. I am more ashamed of such a passage than I should be of palpable octaves and fifths, which anyhow are no hindrance to pure intonation.[8]

The article in *Grove* thus launches us on a far-reaching inquiry, adequate discussion of which would carry us beyond the limits of an article in the *Music Review*. But interesting light is thrown on some aspects of the matter if we return to where we began, with the editorial comment in the modern *Grove*, on the impossibility of a return to mean-tone tuning. For though it is idle to try to put the clock back, mean-tone tuning was a practical approximation to the scale system which can be distilled from the music of great masters from Taverner to Handel, and it has much to tell us about the essential problems of all tunings of keyboard instruments.

Let us begin, however, not with origins, but by considering mean-tone tuning as it was understood by knowledgeable musicians in this country little more than a century ago, when it was in practically universal use for the organ and the English piano. Indeed, it survived on the organ until well into the second half of the nineteenth century, and as an organ-tuning we naturally think of it today. It was used on Handel's organ. What then, exactly, was this mean-tone tuning? In the absence of an organ tuned in the mean-tone system on which to discover its good points and its limitations, perhaps the simplest answer is given by a picture[9] showing the nature of this tuning graphically, as in Fig. 2.

In this figure the intervals produced by mean-tone tuning

[8] The passage is quoted in full, with the musical example, in the final essay in *The Musical Ear* [Ref. 11].

[9] The writer is indebted to the Oxford University Press for permitting the use of this figure and Fig. 7, which are adapted from his *Musical Slide-Rule* [Ref. 11], and the reproduction of an illustration from his *Music and Sound* [Ref. 1] which appears as Fig. 4.

[Some recordings are now available, *e.g.* the Compenius organ at Frederiksborg in Denmark.]

are drawn, with mathematical accuracy, on the left. On the right are shown, with a theoretical accuracy to correspond, all the *diatonic intervals* less than an octave, together with the octave itself, the major ninth, and the major tenth, each measured from 'zero'. We may conveniently think of diatonic intervals as being indicated, as nearly as may be, by the white keys of the keyboard; *e.g.* of the minor sixth as the interval measured upwards from E to C, of the tritone as that from F to B, of the imperfect fifth as that from B to F, and of the minor seventh as that from G to F, or D to C, all measured upwards. The sizes given to the fifth, the fourth, and the thirds and sixths, in Fig. 2, are those they would have in the appropriate concord of sixteenth-century counterpoint if sung perfectly in tune. What of the remaining diatonic intervals? Let us experiment. Take a slip of stiff note-paper, and lay it on the left-hand side of Fig. 2 with its edge touching the centre line. Make marks on it, with a sharp pencil, opposite the zero, the fourth, and the fifth, as accurately as possible. Now slide the paper slip up, till the zero mark is opposite the *major* (greater) tone. The mark for the fourth will be opposite the fifth. The major tone is now seen as the interval between C and the dominant of G.

Now slide the paper slip up, till the zero mark on it is opposite the major third. The fourth marked on it will be opposite the major sixth. (The musical tones corresponding to the zero, the major third, and the major sixth, on the right-hand side of the figure will produce a 6/3 concord.) Now slide the paper slip down till the fifth marked on it is opposite the major sixth. The zero mark is now opposite the *minor* (lesser) tone. The minor tone is here seen as the difference between a major sixth and a fifth. It is also the interval we must add to a major tone to build up a major third (try it with the paper slip and see). That is why the musical scale has two kinds of whole tone for true concords produced on instruments of free intonation, such as stringed instruments. Moreover, these two kinds of whole tone are sometimes inter-changed, *e.g.* to produce a perfect fifth between the second- and sixth-degree notes of the scale, as we have just seen. We have now obtained all the information we need to understand mean-tone tuning.

Obviously the organ-builder cannot provide notes which move about, to do duty at one time in a major tone, at another in a minor tone. Thus, the first thing the organ-tuner has to do is to discover a workable compromise. Mean-tone tuning

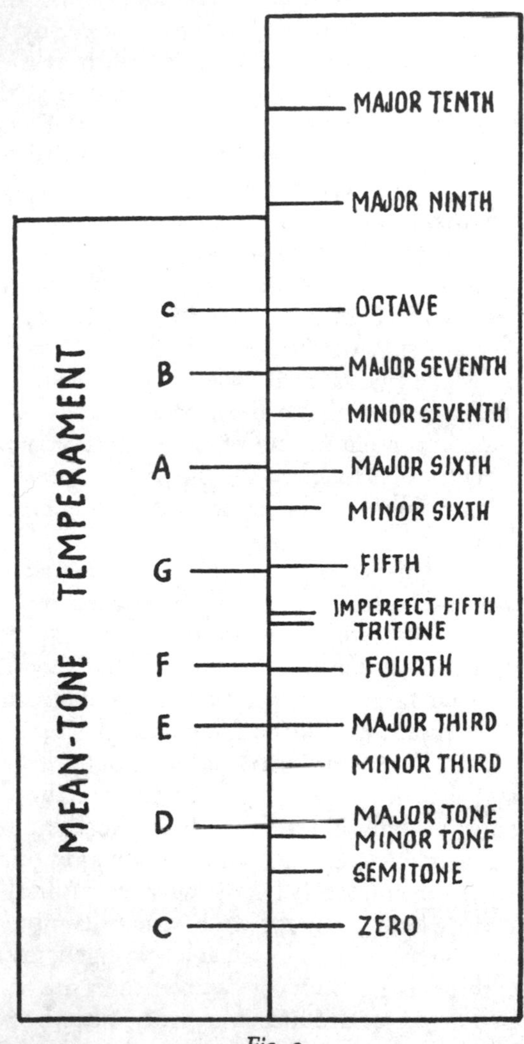

Fig. 2

adopted the simple plan of splitting the difference between the major tone and the minor tone: whence its name. It follows, as Fig. 2 shows, that the major third, which is the sum of a major and a minor tone, is tuned perfectly true in mean-tone temperament. How shall we tune the fifth?

Turn back a couple of paragraphs. The interval by which a major sixth exceeds a *minor* tone is a fifth. Consequently that by which a major sixth exceeds a *major* tone is less than a fifth. It is too small by the difference between a major and a minor tone, a small theoretical interval called a comma. In the scale of C, we may conveniently think of the major tone as giving us D, and of the major sixth as giving us A, each being measured from C; but between D and A we should have an interval which is a flattened fifth. To make a true fifth with D as the dominant of G we should have to sharpen the A by a comma. In turn we should have to sharpen the E above it by a comma; for the interval by which a major tenth exceeds a major sixth is a perfect fifth. (Try it on Fig. 2 with the fifth which you marked on your paper slip.)

We began, four paragraphs back, by finding D as the dominant of G, which in turn is the dominant of C. Thus, beginning with C, we have in effect now tuned four successive fifths perfectly in tune: CG, GD, DA, and AE; and we have reached an E that is a comma sharp. But in mean-tone tuning we make all major thirds true. The solution of the tuner's problem is obvious: split the superfluous comma among the four fifths, and flatten each one by ¼ comma. G will be ¼ comma flat, D will be ½ comma flat. But A will not be ¾ comma flat. A true DA is a comma less than a true fifth. Therefore A, as now tuned, will come out ¼ comma sharp. In turn this will give an E that is perfectly true, two octaves and a major third above the C we began with.

Since a mean-tone fifth is ¼ comma flat, a mean-tone fourth, required to complete the octave, is ¼ comma sharp. This fixes F for us, a mean-tone fourth above C. It remains to fix the tuning for B, as the major third of the mean-tone dominant, G, in a *full close*. Obviously, since G is ¼ comma flat and all major thirds are true in mean-tone tuning, B will also be ¼ comma flat. The mean-tone semitone Bc, like that between E and F, is ¼ comma larger than the true diatonic semitone. We have now tuned all the white keys, for the scale of C, in mean-tone temperament. Examine the result on the left-hand side of Fig. 2, in which the width of all the black graduation marks was made, as nearly as possible, that representing ¼ comma. In practice, the tuning was accomplished by tuning the notes of an octave, or so, by successive fifths tuned upwards as mean-tone fifths, a mean-tone fourth, tuned downwards, being interpolated, as required, to restrict the compass of the notes used, as Lecky explained in the original article in *Grove*. The right degree of tempering was determined by the rate of beating; and, for reasons explained in books on musical acoustics, a fourth tuned downwards from a given note in any temperament produces beats at the same rate as a fifth tuned upwards from the same note in the same temperament.[10] The tuner checked his results by trial with the 'sweet' major thirds. This procedure made use of the black keys of the keyboard, to which we now turn. The left-hand side of Fig. 7, below, shows the tuning of those keys.

For a limited range of key we tune our five black keys to give the same intervals as the white ones. We so tune F♯, C♯ and G♯, in the manner explained above, that severally they produce true major thirds with the mean-tone dominants of the keys of G, D, and A respectively. We have two black keys left. One we will tune as a mean-tone fifth below F (or a mean-tone fourth above it, which is the same thing), giving us B♭, the subdominant of F, a mean-tone semitone above A. The other we will tune as a mean-tone fifth below B♭, giving us E♭, the subdominant of B♭, a mean-tone semitone above D. For if we add a mean-tone semitone to a mean-tone fifth the quarter commas of the tempered intervals cancel each other, and we obtain a true minor sixth. As this is the complement of a true major third, required to make up an octave, mean-tone temperament produces minor sixths, such as AF or DB♭, that are true. All our black keys are now used up, but we have no D♭, D♯, G♭, A♭, or A♯. If we try to use C♯, E♭, F♯, G♯, or B♭ instead, we shall produce what our great grandfathers called a 'wolf' (because of its howling). Why is this?

[10] This was verified experimentally, and deduced from his theory of beats, by Dr. Robert Smith, F.R.S., as set out in his *Harmonics* (2nd ed. 1758) in a passage, p. 93, under the heading 'To show that the theory of beats agrees with experiments': see a paper by the present writer in *Phil. Mag.*, XXXIV (7), p. 472 (July, 1943) [Ref. 27].

On the right-hand side of Fig. 2 we have two diatonic intervals nearly the same: the tritone and the imperfect fifth. If the zero is C these intervals become CF♯, the tritone of the key of G, and CG♭, the imperfect fifth of the key of D♭, as shown in musical notation in Fig. 3. If we assume these discords

Fig. 3

to be tuned perfectly true (say by reference to the respective dominants)[11] they fix F♯ and G♭ for a theoretically exact tuning with true intonation.

In Fig. 4 the relevant intervals, from the fourth to the fifth as shown in Fig. 2, are redrawn on a much larger scale; and to redraw them use has been made of the fact that there are almost exactly 9½ commas in a major tone, and 5·2 commas in a diatonic semitone (a comma being, as noted above, the difference between a major and minor tone). The intervals FG♭ and F♯G, being diatonic semitones, are each 5·2 commas and therefore the interval F♯ G♭, by which they overlap, is

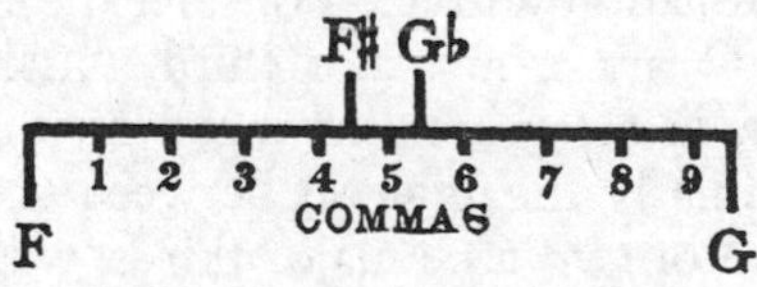

Fig. 4

0·9 commas (these arithmetical niceties assume a theoretically exact tuning). Now let us go back to mean-tone tuning, and let us suppose that our keyboard has additional black keys for the five 'black notes' which we found, in the previous paragraph, to be lacking in mean-tone tuning. A mean tone is ½ comma less than a major tone, *i.e.* it is equivalent to only 9 commas. On the other hand, a mean-tone semitone, being

[11] To tune down a major tone from the dominant we should tune down a perfect fifth and then up a perfect fourth.

¼ comma larger than a diatonic semitone, is equivalent to 5·45 commas. On both accounts F♯ is flattened and G♭ would be sharpened in mean-tone tuning. The result is that, in mean-tone tuning, the intervals FG♭ and F♯G overlap much more than they would if tuned with true intonation (Fig. 4); and F♯ and G♭ are now 1·9 commas apart—which is more than twice the corresponding difference with true intonation, or more than one-third of a diatonic semitone, or nearly 8 times the error in the mean-tone fifth. This result is exhibited in Fig. 5. Even the most tolerant ear boggles at the beating (the wolf's howling) produced by using a mean-tone F♯ for G♭.

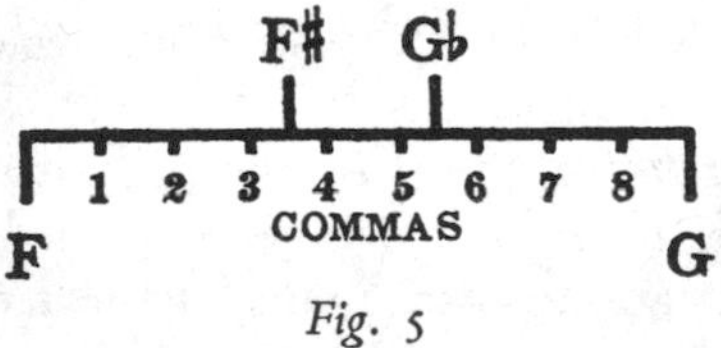

Fig. 5

Since all whole tones are the same size in mean-tone tuning, the mean-tone interval between each pair of adjacent sharps and flats will be the same size as that shown in Fig. 5, for F♯ G♭ in mean-tone tuning. It is interesting to discover that this interval is the same as the interval G♯ A♭ if both notes are tuned in true intonation with C. For since the interval between a true A♭ and C is a true major third, this interval will be exactly the same in mean-tone tuning, which makes all major thirds true. Similarly the interval between C and a true G♯, which is the sum of two true major thirds, will be exactly the same in mean-tone tuning. This is illustrated by Fig. 6, which gives the positions of G, G♯, A♭, and A, which are required for theoretically true intonation with the tonic, C. The interval G♯ A♭ is the same as the interval F♯ G♭ in Fig. 5, but the interval GA is ½ comma less than the mean tone in

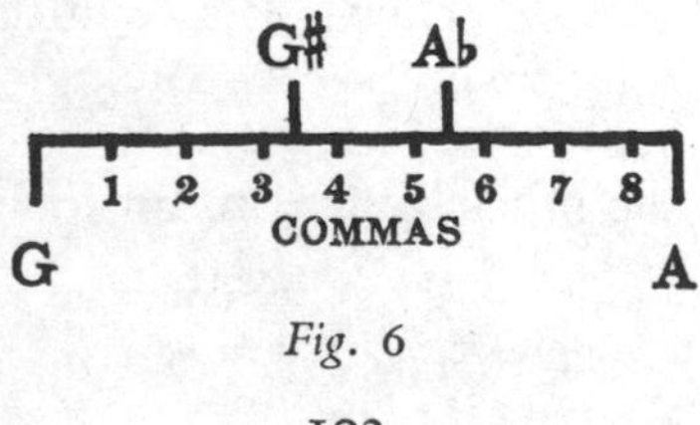

Fig. 6

Fig. 5, because it is a minor tone, as we saw earlier in this chapter.

It is evident that in mean-tone tuning we must be content with only five chromatically altered notes, and the available good keys are limited to the major keys of C, G, D, A, F, and B♭, and the minor keys of G, D, and A. It was perhaps the limited number of good minor keys, in particular, that prompted the builders of some organs to provide two more pipes to the octave for each organ-stop, so as to give them D♯ as well as E♭, and A♭ as well as G♯. Each of the two corresponding black digitals was divided, transversely, so as to do duty for two digitals—one half playing the flat and the other the sharp[12] (see Fig. 7 below). This added E minor and C minor to the available minor keys, and E major and E♭ major to the available major ones. For sustained notes, such as those of hymn-tunes, the device would not be too inconvenient.

To sum up: the fifths and fourths were more mistuned than in equal temperament but, as experience showed, not to an extent which offended the organist's ear. The major thirds were true, and the minor ones had less than half the error of the equally tempered minor third. Yet despite the harmonious effect of this tuning, its limitations of key and its inability to provide for much chromatic harmony led to its abandonment with changing taste in Church music. The differences between the two temperaments are shown graphically in Fig. 7.

If, however, we travel back to an earlier period we discover why it was that, for a keyboard instrument with fixed intonation, such as the organ, mean-tone temperament was the best possible approximation to the flexible intonation of the polyphonic period. As an example, consider the opening strain of Taverner's *Western Wynde* Mass, written before mean-tone

[12] This device was adopted as early as the sixteenth century, for Salinas records that he found it on an organ in Florence (*vide* Grove's *Dictionary of Music*, 1st ed. Vol. IV, p. 73). *See also* Chapter 6, p. 54. It was referred to by Pier. Francesco Tosi, in his *Observations on the Florid Song*, written in Italian and published in Bologna in 1723 (English trans. 1743). It was dismissed by Dr. Robert Smith in his *Harmonics* (2nd ed. 1758) with the comment given on p. 60 Chapter 7. Yet the vogue of mean-tone tuning for organs lasted, without it, for another century.

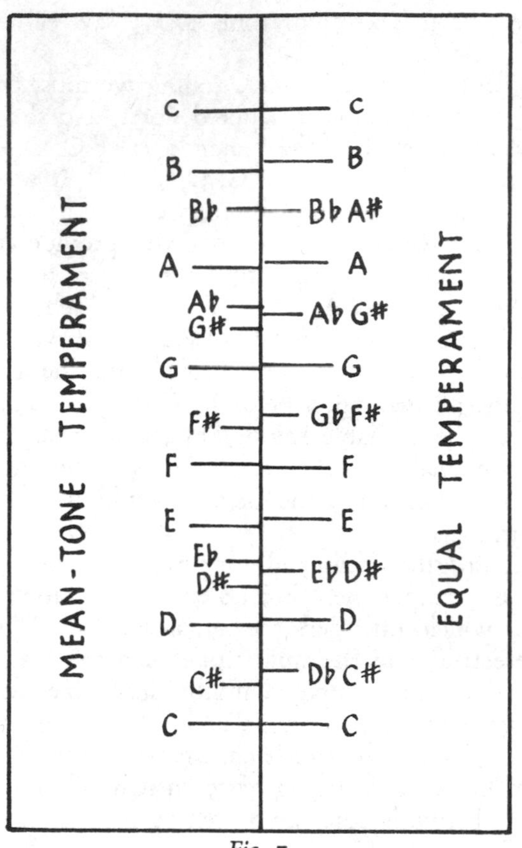

Fig. 7

tuning was standardised by Salinas. This is shown in Fig. 8.[13] In an article entitled 'The Musical Scale' in the issue of the *Musical Quarterly* for April, 1942 [Ref. 26], the present writer analysed the various concords in the passage and showed that, *as heard* by a musical ear, not less than four of the notes written in it must be mutable, *viz.*:—G, C, E, and A, if the concords are all to sound perfectly in tune. The scale-system of modal polyphony was surprisingly flexible, and it is quite evident

[13] The writer is indebted to the Carnegie United Kingdom Trust and the Oxford University Press for permission to transcribe this example from Vol. I of *Tudor Church Music* (he has inserted modern barring). It also appears on p. 44.

[See also the edition by Brett published by Stainer & Bell Ltd.]

that, in this example, the degree of flexibility is connected with the absence of the tonality of a later period, for to the modern ear the music sounds at first as if it alternated between the keys of G minor and F major and occasionally touched B♭ major and D minor.

Fig. 8

It is easy to see from this example why the device of a mean tone on a keyboard instrument, adopted to serve as either a major or a minor tone, commended mean-tone tuning in the sixteenth century as the best way of coping with these mutable notes and of approximating to the flexible scale-system as well as was possible on an instrument of fixed intonation. And this example also explains why C♯, E♭, F♯, G♯, and B♭ were chosen for the 'black' keys of the keyboard. Taverner's Mass was written in the Dorian mode transposed, and it makes use of B♭, E♭, and F♯, as well as B♮, E♮, and F♮. An authentic cadence on D, in the Dorian mode proper, required C♯, while a similar cadence on A, in the Aeolian mode, required G♯. As R. O. Morris (1886–1948) points out in his *Contrapuntal Technique in the Sixteenth Century*, p. 11, 'these five chromatic notes B♭, E♭, F♯, C♯, G♯ are the only chromatic notes permitted in the strict modal system of the sixteenth century', and they are precisely those given by mean-tone temperament.

The noteworthy thing about mean-tone tuning is that, taking its origin in modal polyphony, before the sense of classical tonality had been developed by composers, it should nevertheless have served, at least for tuning the organ, whose intonation is more insistent to the ear than that of the piano, for some four hundred years; for a long time, that is, after classical tonality had attained its dominating influence on later music. This historical fact is of considerable significance in estimating the influence, if any, which temperaments may have exercised on the composer's mode of thought. Unless we take account of mean-tone temperament as well as equal temperament we are failing to take account of all the relevant evidence. That of itself is a sound reason for trying to understand this now forgotten tuning, not, in textbook fashion, as a technical problem for the tuner, but as it appealed to the musical ear, which is what we have been attempting to do.

When we take both temperaments into account it is evident that we must think, separately, of two different aspects of the matter. The first is the employment of a rigid intonation, as measured by physical vibrations, inseparable from any keyboard music *as played* (not necessarily the same thing *as heard*). The second is the distortion of what should be consonant intervals, and in particular those of the concords of sixteenth-century polyphony, which is characteristic of one or other temperament. For it is in this second characteristic that the two temperaments differ.

Consider, first, the rigid intonation which, for 600 years, the keyboard has tried to impose on our ears, and on the musical scale used by composers. Here the acid test is the intonation of the string quartet, and that of a good cathedral choir trained to sing unaccompanied. Even Ellis, making use of A. J. Hipkins' very accurate ear, recorded that the flexible intonation of strings and voices was an established fact. Three-quarters of a century ago he wrote[14]:

> The want of fixed tones both on the bowed instruments and the human voice, and the extreme ease with which pitch can be almost involuntarily and unconsciously altered to suit the feeling or circumstances of the moment, while forming of course the great point of perfection which distinguishes these musical instruments from all others, yet occasion great difficulties in the way of scientific investigation. It is impossible to depend with certainty on hearing the same intonation twice.

But surely, if the flexible intonation of a good string quartet, or a good unaccompanied choir was really their great point of musical perfection, this observation is conclusive. It demolishes

[14] *Sensations of Tone*, 1st ed., p.483. (See also *Note* on p.111.)

at a stroke Ellis' *a priori* theory that composers whose writing for strings and voices discloses, in fact, the use of mutable notes, for true concords, thought in tempered music. His 'great difficulties' arose, not from scientific investigation, but from his attempt to confine the practice of artists, and the procedure of composers, to the conclusions he reached from the rigid intonation of his experimental harmonium, conclusions which he embodied in the duodenarium he devised to guide composers in restricting their freedom of modulation in the way he thought desirable. We are reminded that Hauptmann's criticism of the piano as an instrument to accompany his violin was not of its faulty intonation but of its rigid intonation. As an illustration he took the string player's practice of playing F♯, as a decorating note, above G♭ (cf. Fig. 4), or C♯ above D♭.

Here is the most important lesson of mean-tone tuning, and it is significant also for equal temperament. Until we use the piano to accompany strings, the rigid intonation of the instrument *as played* matters less, and must always have mattered less, than the *a priori* theorist supposed. To understand why, we must turn to the properties of ear and the brain, and consider the intonation of the piano *as heard*, particularly by those who think contrapuntally. For the *a priori* theorist assumed that the intonation we hear corresponded, exactly, to the vibrations we listen to, whereas the intonation we hear 'is correlated directly only with the neural activities occurring in that portion of the brain which is known as the cerebral cortex'.[15]

Like mean-tone tuning, equal temperament adopts a compromise whole tone, which, as Fig. 7 suggests, is about a third of a comma larger than a mean tone. The characteristic which really distinguishes equal temperament from mean-tone temperament is the sharp major third and the 'narrow' minor third, with corresponding modification in the sixths. This is particularly evident on the organ when played loudly, with the more penetrating reeds and the mutation stops and mixtures. These strengthen the higher partial tones of the diapason and give increased definition to its notes and to the intervals between them. Even then the harshness of the thirds

[15] The *Music Review*, Vol. III, No. 2, p. 100.

may be mitigated by one of the most important stops of the instrument, the acoustic quality of the building. The extreme case is the harmonium on which equal temperament is almost intolerable. There was much to be said for the efforts of the nineteenth-century theorists to provide it with so many notes to the octave that its cost would be prohibitive, and to equip it with a complicated keyboard which would make it unplayable.

At this point our growing knowledge of the science of hearing again takes a hand. There is great variation in the acuity of hearing of different pairs of ears. That acuity may be strengthened by careful training, or damaged by inattention and carelessness.

No doubt some composers have been influenced by the actual intonation of the piano, particularly in dealing with chromatic passages or in using a chromatic idiom. In so far as this is true the evidence must be found in their music, for example, in certain passages which it is difficult to sing or play in tune: mere conjecture is not good enough. For to sing or play with any accuracy in equal temperament is, acoustically, a far more difficult thing than the theorist imagines. To a sensitive ear, listening intently, a consonance is sharply defined while a dissonance, which includes a tempered, *i.e.* a mistuned, consonance, lacks definition. The ear, so to speak, can only make a shot at equal temperament. For many pianists the sharp major third of their instrument may possibly induce a vague kind of listening which helps the ear to accept the mistuned intervals as good enough and the rigid intonation as a musical scale. On the other hand, there are well authenticated instances of organists who have come to prefer the faulty thirds of their instrument and criticise their choirs for singing thirds, unaccompanied, which sound flat to them. But how many pianists have so far absorbed the tuning of their instrument that they criticise the intonation of a good string quartet? On the contrary, most pianists find it particularly beautiful.

It is beyond question that *a priori* theory plays too large a part in the notion that composers think, indeed *can* think in any exact sense, in equal temperament. With all its logical implications this *a priori* theory proves too much. If a composer is obliged to think in terms of the intonation of his instruments, we must suppose that Bach had to think in the 'pure scale' of polyphony, with its flexible intonation, when writing for his choir, in mean-tone tuning when writing for his organ, and in something like equal temperament when writing for his clavichord. A theory which proves too much stands self-condemned. Is it not simpler, and more reasonable, to imagine Bach thinking, like any serious student of counterpoint today, in terms of musical intervals? Enharmonic modulation is the particular trouble of the *a priori* theorist who assumes (for it is pure assumption on his part) that it implies equal temperament, which identifies F♯ with G♭, and so on. Tovey is insistent that enharmonic modulation implies true intonation and a flexible scale (he writes of the 'unstable' intonation of mutable notes); for otherwise it must lose its musical significance as a 'sublime mystery'. To assign it to equal temperament is surely to put the cart before the horse. Equal temperament is the best approximation, on an instrument of fixed intonation, to the flexible intonation implied in enharmonic change. Those writers who find in equal temperament an explanation of the procedures of modern music must undertake the forbidding task of convincing us that musical ears, today, have lost qualities they certainly possessed for many centuries—qualities which made them conscious of enharmony.

We must be careful to except those composers whose protagonists claim that their atonal music is breaking new ground by the use of a scale which implies a closer approximation than the piano-tuner can achieve, in practice, to the division of the octave into twelve exactly equal intervals. That is a deliberately *conscious* effort and it demands a new technique. Indeed, were the deliberate use of this intonation the whole story, one would be inclined on scientific grounds to postulate a technique which, on all possible occasions, contradicted that of the sixteenth century, as being an experimental attempt to defeat the natural properties of the ear. What we are concerned with, however, is the *unconscious* effect, if any, of equal temperament on the procedures of other musical composition, particularly that of composers who have abandoned the nineteenth-century academic practice of giving harmony pride of place over polyphonic writing and the study of counterpoint, and who have not sought to banish pure consonance from their vocabulary. There is some ground for thinking that temperament has long ago done its worst and that, in the main, musical composition has shied away from its rigid intonation.

The Editor of *Grove*, in the note from which we quoted at the outset, expressed the view that since Lecky wrote his article, 'composers . . . have increasingly tended to think in terms of Equal Temperament, witness the whole-tone scale and other developments of Harmony'. With great respect for his musical scholarship, we may doubt whether this does not show too much deference to the 'theorist'. The reference to the whole-tone scale is significant. In 'theory' this scale presupposes the use of every alternate note in an octave of twelve semitones of equal temperament. But 'theory' and practice are very different things; and, unless it is the fruit of musical scholarship, musical theory is usually most unscientific in its method and outlook. Tovey expressed the opinion that the whole-tone scale had far less to do with Debussy's music than is commonly supposed.[16] Here again the acid test is the intonation of the string quartet. The slow movement of Debussy's quartet has some beautiful phrases which owe their charm precisely to the fact that they are played perfectly in tune. Take, as a typical example, the phrase (Fig. 9) quoted

Fig. 9

by Edwin Evans in his article on Debussy in Cobbett's *Cyclopedic Survey of Chamber Music* with the comment: 'It would surely need no more than this one bar to identify the composer!'[17] Now look at it acoustically, which means

[16] Brief quotation is impossible. Readers will find the witty original in the article 'Harmony', *Enc. Brit.*, 14th ed.

[17] The phrase is here transcribed by kind permission of the Oxford University Press.

in terms of the natural properties of the ear. Observe the perfect fifths which build into the harmonic structure, and recall that the fifth has very sharp definition and insists on true intonation. Look at the diatonic intervals of the first violin and viola, and at the scale in the cello part, ending by moving from a minor sixth to a major sixth below G♭. Even the sequence of falling semi-tones in the second violin part is a melodic device as old as the hills, though the player is here helped over two difficult stiles by perfect fifths with the first violin. Only in their last notes do the middle parts meet a surprise; and the cello, aided by the first violin, determines their notes by a pure concord. Granted a string quartet that observes Tertis' direction to listen *intently*, it would be hard to find a passage which, despite its elusive tonality, would appear to be more difficult to play in equal temperament instead of perfectly in tune. It remains true today that, as Clerk Maxwell said[18] nearly three-quarters of a century ago,

> The special educational value of this combined study of music and acoustics is that more than almost any other study it involves a continual appeal to what we must observe for ourselves,

for this is the touchstone of the scientific method, which insists on bringing all conjecture, such as that of the musical 'theorist', to the test of experimental investigation and accurate observation.

[18] In the Rede Lecture at Cambridge, 1878.

[*Note*. A footnote by Hipkin's daughter to the article on Ellis in *Grove III & IV* reads as follows:—'Ellis's exhaustive experiments were made entirely by calculation, as he was tone-deaf and unable to distinguish one tone or tune from another. Hipkins tested each experiment by ear.']

BIBLIOGRAPHY OF THE WRITINGS OF LL. S. LLOYD

(1) *Music and Sound*, pp. 181; Oxford University Press, 1937.

(2) Decibels and Phons for Musicians, *Musical Times*, Vol. 78, No. 1138, pp. 1023–5. December 1937.

(3) *A Musical Slide-Rule*, pp. 25; Oxford University Press, 1938.

(4) *Decibels and Phons: A Musical Analogy*, pp. 18; Oxford University Press, 1938.

(5) Electronic Organs and the Phonodeik, *Musical Times*, Vol. 79, No. 1147, pp. 682–5. September 1938. (Chapter II of *The Musical Ear*.)

(6) Intonation—and the Ear, *Music and Letters*, Vol. XIX, No. 4, pp. 443–449. October 1938. (Chapter I of *The Musical Ear*.)

(7) The Sound of Church Bells, *Musical Opinion*, Vol. 62, No. 736; January 1939. Vol. 62, No. 737; February 1939. Vol. 62, No. 738; March 1939. (Chapter IV of *The Musical Ear*.)

(8) Helmholtz and the Musical Ear, *Musical Quarterly*, Vol. XXV, No. 2, pp. 167–175; April 1939. (Chapter III of *The Musical Ear*.)

(9) Just Temperament, *Music and Letters*, Vol. XX, No. 4, pp. 365–373; October 1939. (Chapter 9 of this book.)

(10) The Place of 'Hearing' in Theory about Music, *Proceedings of the Royal Musical Association*, 66th Session, pp. 33–52. 1939/40.

(11) *The Musical Ear*, pp. 87 (includes refs. (5), (6), (7) and (8)); Oxford University Press, 1940.

(12) A Note on Just Intonation, *Journal of the Acoustical Society of America*, Vol. 11, pp. 440–445 (line 4, para. 1, col. 1, p. 441 should read 'The consequence was *un*fortunate . . .); April 1940.

(13) The Perfect Fifth, *Musical Times*, Vol. 81, No. 1169, pp. 298–300; July 1940. (Chapter 1 of this book.)

(14) The Major Third, *Musical Times*, Vol. 81, No. 1171, pp. 365–367; September 1940. (Chapter 2 of this book.)

(15) The Myth of Equal Temperament, *Music and Letters*, Vol. XXI, No. 4, pp. 347–361; October 1940. (Chapter 8 of this book.)

(16) A Forgotten Tuning, *Musical Opinion*, Vol. 63, No. 746, pp. 58–59; November 1940.

(17) The Harmonic Series, *Monthly Musical Record*, Vol. 71, No. 824, pp. 35–39; February 1941.

(18) The Perception of Pitch, *Musical Times*, Vol. 82, No. 1178, pp. 140–141; April 1941.

(19) The Artificial Scale, *Musical Times*, Vol. 82, No. 1181, pp. 252-254; July 1941. (Chapter 4 of this book.)

(20) Musical Theory in Retrospect, *Journal of the Acoustical Society of America*, Vol. 13, pp. 56–62; July 1941.

(21) Musical Theory in the Early Philosophical Transactions, *Notes, Royal Society*, Vol. III, 1940-41.

(22) Musical Theory in the Making, *Monthly Musical Record*, Vol. 71, No. 832, pp. 226–231; December 1941.

(23) The Musical Scale, *Musical Quarterly*, Vol. XXVIII, No. 2, pp. 205–215; April 1942.

(24) Speculative Music, *Music Review*, Vol. III, No. 2, pp. 921-02; May 1942. (On p. 91 of this issue is printed a note by Ll.S.Ll. in answer to 'Theorists in the Dark' by Noel Heath Taylor—an article criticising his 'The Myth of Equal Temperament'. See (15) above.)

(25) Modern Science and Musical Theory, *Journal of the Royal Society of Arts*, Vol. XC, No. 4619, pp. 581–594; August 1942.

(26) The Perception of Small Intervals and Beats, *Musical Times*, Vol. 84, No. 1201, pp. 75–77; March 1943. (Chapter 3 of this book.)

(27) Just Intonation Misconceived, *Music and Letters*, Vol. XXIV, No. 3, pp. 133–144; July 1943.

(28) The Problem of the Keyboard Instrument, *Philosophical Magazine*, Vol. XXXIV, 7th Series, No. 234, pp. 472–479; July 1943. Vol. XXXIV, 7th Series, No. 236, pp. 624–631; August 1943. Vol. XXXIV, 7th Series, No. 237, pp. 674–684; October 1943.

(29) Pseudo-Science in Musical 'Theory', *Proceedings of the Royal Musical Association*, 70th Session, pp. 35–51; 1943/44.

(30) Lemuel Gulliver and Musical Theory, *Music Review*, Vol. 1, No. 1, pp. 40–45; February 1944.

(31) The Lesson of Mean-Tone Tuning, *Music Review*, Vol. V, No. 4, pp. 214–227; November 1944. (Chapter 10 of this book.)

(32) The *a Priori* Theorist and Music, *Music and Letters*, Vol. XXVI, No. 2, pp. 97–102; April 1945.

(33) Recent Research in Piano-Tuning, *Musical Opinion*, Vol. 68, No. 814, pp. 295–296; July 1945.

(34) Hornbostel's Theory of Blown Fifths, *Monthly Musical Record*, Vol. 76, No. 873, pp. 3–6; January 1946. No. 874, pp. 35–38; February 1946.

(35) The Myth of Equal-Stepped Scales in Primitive Music, *Music and Letters*, Vol. XXVII, No. 2, pp. 73–79; April 1946.

(36) Notes or Tones?—a Lost Opportunity, *Monthly Musical Record*, Vol. 76, No. 881, pp. 203–207; November 1946.

(37) Concerning 'Theoreticians' and Others, *Music Review*, Vol. VIII, No. 3, pp. 204–213; August 1947.

(38) The Loudness of Pure Tones, *Musical Quarterly*, Vol. XXXIII, No. 4, pp. 481–489; October 1947.

(39) Temperament—Without Tears, *Monthly Musical Record*, Vol. 78, No. 897, pp. 123–127; June 1948. (Chapter 6 of this book.)

(40) International Standard Pitch, *Journal of the Royal Society of Arts*, Vol. XCIII, No. 4810, pp. 74–89; December 1949.

(41) British Standard Musical Pitch, *Music Review*, Vol. II, No. 2, pp. 109–117; May 1950. (A lecture delivered to the Acoustics Group of the Physical Society on 10th November, 1949, under the title 'Concert Pitch'.)

(42) Equal Temperament, *Monthly Musical Record*, Vol. 80, No. 917, pp. 118–123; June 1950. (Chapter 7 of this book.)

(43) The Bagpipe Scale, *Monthly Musical Record*, Vol. 71, No. 824, pp. 237–240; November 1950.

(44) What are Phons? *Musical Times*, Vol. 93, No. 1307, pp. 18–19; January 1952. Vol. 93, No. 1308, pp. 62–63; February 1952.

(45) The Loudness of Musical Tones, *Music Review*, Vol. XIII, No. 2, pp. 101–109; May 1952.

(46) The Sensation of Loudness, *Music and Letters*, Vol. XXXIX, No. 3, pp. 243–250; July 1953.

(47) The History of our Scale, *Music Review*, Vol. XIV, No. 3, pp. 173–185; August 1953. (Chapter 5 of this book.)

(48) The Strike-Notes of Church Bells, *Music and Letters*, Vol. XXXV, No. 3, pp. 227–232 and 240; July 1954.

(49) Conflicting Musical Theories of the Nineteenth Century, *Monthly Musical Record*, Vol. 85, No. 967, pp. 119–122; June 1955.

In addition to the above, Ll. S. Lloyd contributed many articles to dictionaries and encyclopedias. His articles in the fifth edition of Grove's *Dictionary of Music and Musicians* are no doubt his greatest contribution to the study of musical acoustics.

PART II

MUSICAL ACOUSTICS

INCLUDING DEFINITIONS, EXPERIMENTS, MEASUREMENTS, CALCULATIONS, VARIOUS APPENDICES AND TABLES

By Hugh Boyle

11

MUSICAL VIBRATIONS

Sound—sound waves—musical vibrations—displacement—cycle—period or *periodicity—frequency—audio-frequencies—standard musical pitch—amplitude—phase—wave—wavelength—transverse wave—sound wave—longitudinal wave—condensations—rarefactions—associated displacement curve—pendular, simple, or harmonic motion, sine wave or curve—free vibrations—natural frequency—damped vibration—maintained vibration—waveform—periodic vibration—fundamental or repetition frequency—harmonic vibrations—harmonic partial vibration or harmonic—inharmonic partial vibration—inharmonicity—forced vibrations—sympathetic vibration or resonance—nodes—antinodes—modes of vibration—sections, loops or ventral segments—coupled system—resonator—false beats—intensity—decibel—threshold of feeling—threshold of hearing—physical beating.*

Sound is, strictly speaking, something heard, but in this type of work its meaning is better restricted to the study of vibrating bodies and vibrations in the air—(*sound waves*)—for which purpose it is commonly used by physicists. When so defined sound is quite independent of the human ear.

All sources of sound vibrate.

A *musical vibration* is a more or less regularly repeated complex to-and-fro motion or *displacement* either side of a position of equilibrium.

Any such single completed to-and-fro motion is termed a *cycle* and the time taken for this to occur is called the *period* or *periodicity* of the vibration.

The *frequency* of a vibration is defined as the number of

cycles which occur in one second. Frequency and period or periodicity are thus reciprocal quantities—the former being measured in cycles per second and the latter in seconds (usually fractions of) per cycle. The maximum frequency range over which sound waves are audible (*audio-frequency range*) varies considerably with individual and with age but may be taken as extending, at its maximum, from about 20 to about 20,000 cycles per second, *i.e.* about 10 'octaves'. Some young children have been found to have a top limit as high as 40,000 cycles per second—however, in all cases it falls continuously with age.

To many who have studied a little physics the terms 'cycle' and 'frequency' may unwittingly have come to be associated *only* with simple harmonic motion (*q.v.*) and its associated sine-wave displacement curve. Such people might do well to re-think these matters in terms of 'periodic time' or 'periodicity' rather than 'frequency'.

'Hertz' is another name for 'cycles per second'.

The frequency of transverse vibration of a uniform string varies directly as the square root of the tension, and inversely as its length and density.

The frequency of vibration of a column of air varies inversely as its length.

The *standard musical pitch* is based on a frequency of 440 cycles per second for A in the treble stave. In equal temperament this amounts to 523·25 cycles per second for the C above it.

The *amplitude* of a vibration is the maximum amount of displacement of the vibrating body or particle from its position of rest or equilibrium.

The *phase* of a vibration is the fraction of its cycle that has occurred, or the fraction of its period that has elapsed, since it last passed through its position of equilibrium in the direction chosen as positive. If any two vibrations of the same frequency do not pass through corresponding positions of their cycles at the same time they are said to be out of phase—or a phase difference is said to exist between them.

Any disturbance which travels through a medium without moving it bodily, and which does so without suffering any appreciable change of form in the process, is called a *wave*.

The distance between any two consecutive vibrating

particles which are in the same phase, or the distance travelled by a sound wave during one complete vibration or cycle, is called its *wavelength*. The wavelength, in air, of middle C is a little over four feet (1·22 metres).

Waves in which the direction of vibration of the particles is at right angles to the direction of travel of the wave are known as *transverse waves*, *e.g.* waves on the surface of a pond.

Sound waves are known as *longitudinal waves*, since, in this case, the vibration of the particles of air takes place along the direction of travel of the wave. The vibrating source of the sound, by producing a series of alternating slight increases (*condensations*) and decreases (*rarefactions*) in the surrounding atmospheric pressure, causes each particle of air to vibrate backwards and forwards in the direction of travel of the wave, thus passing on the disturbance to its adjacent particles, which then, in turn, reproduce the same disturbance a little later in time—rather like the succession of collisions that occur between loosely coupled railway trucks during shunting operations.

For the study and comparison of sound-waves it is usual to obtain graphs showing the manner in which their displacement (and/or pressure) varies with time. In order to do this the to-and-fro motion (displacement) of the particles of air along the line of travel of the wave must be made to appear as an up-and-down motion, *i.e*, as motion at right angles to the line of travel of the wave. This is most conveniently done electrically by means of a microphone and an oscilloscope—the latter being used to display the graph or curve. Under these circumstances, however, it must be remembered that the trace cannot portray an actual displacement curve of the sound-wave, though it does give a diagrammatical representation from which the physical proportions and properties of the actual sound-wave may be determined. Such a curve is called an *associated displacement curve*. (See p. 26.)

The simplest type of displacement curve is that of the air-vibration from a tuning fork with its stem stuck into a resonance box. Such a vibration is smooth and steady like that of a pendulum swinging through so short a distance that its motion is indistinguishable from motion along a straight line. Ohm and Helmholtz called this type of vibration '*pendular*'. From a mathematical point of view it is the simplest form of motion that a particle can have and is therefore often referred to as '*simple*' or as '*simple harmonic motion*' ('harmonic' here has no connection with 'harmony'). Such a curve can be readily reproduced with the aid of a table of sines—hence it is also sometimes referred to as a *sine wave* or *curve*.

Vibrations such as those from a struck string are called *free vibrations* because the string, after being struck, is allowed to vibrate freely at its *natural frequency*, *i.e.* its motion is not interfered with from outside. Under these circumstances the vibratory motion is gradually destroyed by dissipation of its energy through the formation of waves, or the resistance of forces of a frictional nature, and is said to be *damped*. Such vibrations are of a more or less transitory nature *i.e.* do not at any time achieve a completely steady state of periodic vibration (see later).

When a body is kept vibrating by some external force which does not impose on it any particular frequency, but keeps the vibration going at more or less its natural frequency, then the vibration is said to be *maintained*. The vibrations from pipes and bowed strings are good examples of maintained vibrations. Ideally, the vibrations of the tuning fork mentioned two paragraphs back should be maintained.

Any vibration in which the form of the associated displacement curve or *waveform* for each cycle is identical—such as one that is maintained—is called a *periodic vibration*. The trace produced by such a vibration on the screen of a cathode-ray oscilloscope is quite stationary. Such vibrations may be simple or complex in waveform.

Thus it is only when man interferes with nature, by blowing or bowing, that a more or less completely periodic vibration is produced.

However complex a periodic vibration may be it can always be regarded as having been built up from a number of simple, component or partial vibrations whose individual frequencies are exact multiples of its *fundamental* or *repetition frequency*, *i.e.* from component or partial vibrations whose individual periods fall into the harmonic series (or what amounts to the same thing, whose individual frequencies fall into an arithmetic series). For this reason, such a vibration, when considered as a whole, is often called an *harmonic vibration*, and its component or partial vibrations—including the fundamental—termed *harmonic partial vibrations* and numbered, for convenience, according to the ratio of their frequency with that corresponding to the first, fundamental or lowest possible partial vibration. *Upper harmonic partial vibrations* (*i.e.* all excepting the fundamental) are sometimes referred to briefly as '*harmonics*'.

In Fig. A of the next chapter sets of harmonic partial vibrations are represented by their corresponding numbers (*i.e.* frequency ratios) spaced out in proportion to their logarithms.

On the other hand, a vibration in which the form of the associated displacement curve for each cycle is not identical—such as a damped or natural vibration—cannot be so regarded. For the individual frequencies of its upper component or partial vibrations cannot be exact multiples of its fundamental or repetition frequency *i.e.* the corresponding periods cannot fall into an harmonic series and the partial vibrations, which are perceived by the ear as inharmonic upper partial tones, are accordingly termed *inharmonic.* Such departure from the harmonic series is called *inharmonicity.* No part of the trace of such a vibration can be made to appear stationary on the screen of an oscilloscope. (See also p. 7 and Chapter 15). The same remarks would apply to the vibrations corresponding to musical notes whose upper partial intervals were those of equal temperament or were sufficiently stretched by inharmonicity.

The vibration ultimately set up in a body, due to the application to it of periodic impulses whose frequency does not coincide with the natural frequency of the body, is called a *forced vibration* (*e.g.* Savart's wheel—a toothed wheel and card). If the forcing frequency lies close to or equals the natural frequency of the body or system then a much larger displacement is caused and the body is said to be in a state of *sympathetic vibration* or *resonance* with the applied periodic force.

All musical instruments are examples of *coupled systems*: *i.e.* a source of sound together with some means of increasing its loudness. One of the simplest forms of such a system consists of a tuning fork mounted on a wooden box having one end open—virtually a pipe closed at one end, called a *resonator*—and tuned to the fundamental frequency of the fork. The increase in loudness is, of course, obtained only at the expense of the duration of the sound. Conversely, the duration can be

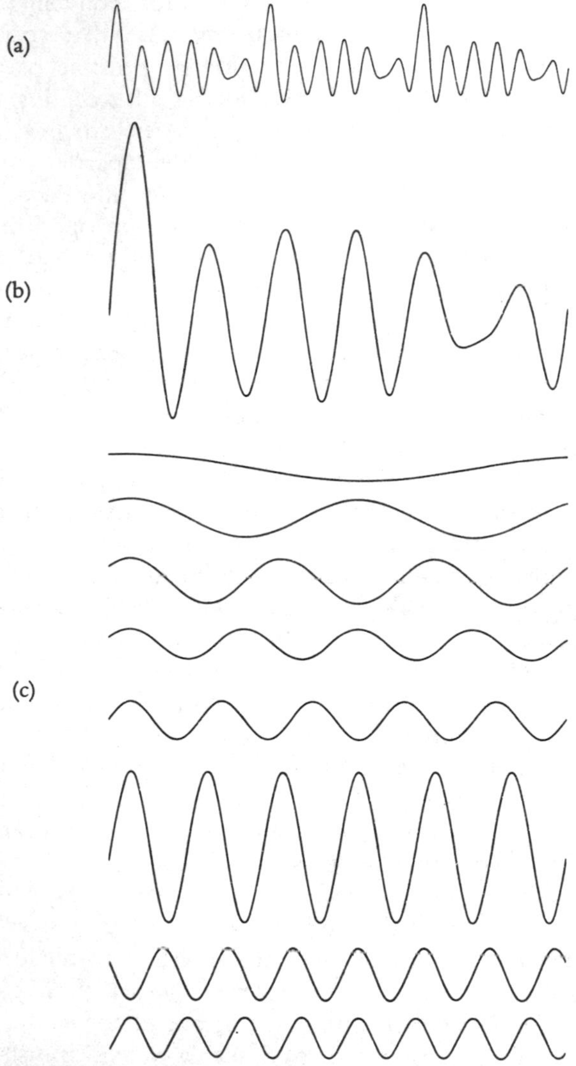

(a) Three cycles of a complex periodic vibration.
(b) One of the complete cycles shown at (a).
(c) The eight harmonic (or simple) periodic vibrations which together make up the single complex periodic vibration shown at (b).

increased somewhat at the expense of the loudness by slightly de-tuning the resonator. (A similar relationship exists between the individual strings of the 'bichords', or 'trichords', of a piano—which are coupled to each other by way of the bridge and soundboard—and the accuracy of the relative tunings of their unisons.) Such de-tuning only affects the frequency of the system very slightly, so that in this case it is almost entirely forced by the natural fundamental frequency of the fork. A further advantage of the resonator is that it more or less eliminates any of the fork's (inharmonic) upper partials which could be present.

Ideally, the bridge and soundboard of any stringed instrument should respond (*i.e.* resonate) equally well to all frequencies and thus 'amplify' the string's vibration without distortion. In practice this condition can never be completely satisfied and the frequency of vibration of the system is somewhat modified by the more prominent resonances in the soundboard's response. (See also *random inharmonicity*, Chapter 15.)

On the other hand, in the reed-pipe coupled system of an organ there is mutual constraint. Neither reed nor pipe is in complete control and both must be tuned to something like the same frequency if the system is to speak satisfactorily.

The air enclosed in any room will be found to resonate to various degrees at many frequencies depending upon its volume and dimensions. In a rectangular room the three lowest of these resonances occur between its three pairs of parallel sides, at wavelengths of half the distance between them—from which figures their corresponding fundamental frequencies can be calcuated (frequency = $\frac{1100}{\text{wavelength}}$ —see Chapter 21). Resonances will also occur at harmonics of these frequencies.

The larger the room, the lower will be these fundamental frequencies, and the greater will be the volume of air required to be moved, hence the greater the time and energy required to build them up and consequently the less objectionable they become.

Again, if the room is made sufficiently large then all these fundamentals and many of their lower harmonics will have such low frequencies that only their higher harmonics will lie in the audio range. Further, the higher the harmonics, the closer together must lie their frequencies, the greater must be the amount of overlapping of their resonance curves, and hence the more uniform must be their response over the audio range. Under these circumstances the room can be regarded more or less as a three-dimensional soundboard.

Points of no apparent motion occurring in a vibrating system are called *nodes*, while those of maximum motion or displacement are called *antinodes*. Nodes are always formed at the fixed ends of vibrating strings, antinodes at the free ends of vibrating air-columns.

By suitably exciting a string or column of air with a continuous simple or pendular vibration of frequency equal to any one of its possible partial vibrations, every particle of the string or air-column can be made to execute simultaneously a simple or pendular vibration about its mean position, and the string or air-column will be found to vibrate in sections formed by alternate nodes and antinodes along its length. Under such conditions the string, or column of air, is said to be vibrating in one of its *modes* (meaning 'manner'—nothing to do with sets of musical intervals) and produces in the ear that partial tone which corresponds with the partial vibration being excited. If, for instance, in the case of a string, the exciting force has a frequency corresponding to—say—the string's 3rd partial vibration, then there will appear along its length 3 antinodes, and, counting in those at the fixed ends, 4 nodes—and so on for frequencies corresponding to other possible partial vibrations of the string (see Fig. G). Under suitable conditions the positions of these nodes and antinodes can be made clearly visible to the naked eye.

The *sections* or '*loops*' between consecutive nodes are sometimes known as *ventral* segments.

A bowed string vibrates, in effect, in all its modes simultaneously and produces a periodic vibration in the air in which the corresponding harmonic partial vibrations may be detected—the number and magnitude of these partial vibrations being determined largely by the position at which the string is bowed. In such a case no well defined antinodes are formed, neither do any nodes become visible, excepting, of course, those forced into existence at the extreme ends by the fundamental physical necessity to limit and clearly define the vibrating length of the string.

The above remarks would be equally true of the vibrations from a struck or plucked string of constant cross-section, stretched between perfectly rigid supports, assuming its motion was undamped, its material perfectly flexible and homogeneous, that the additional tension caused by deflecting it was negligible compared with the tension it was already under in its position of equilibrium, and that the effects of gravity could be ignored. In practice none of these conditions can ever be fully realised, and, to the extent of their deviation from these ideals, all plucked or struck strings must suffer some degree of inharmonicity in their vibrations.

Beats which occur in a *single* string between its lower partial vibrations, due to irregularities in its mechanical properties, are known as *false beats*.

Other things being equal, the longer the string the greater its flexibility, the lower its degree of inharmonicity, and the more pleasing and definite is the note perceived from it.

The rate of flow of the mechanical energy or the power of a sound vibration at any point is called its *intensity* at that point.

The intensity of a simple or pendular vibration is directly proportional to the square of the product of its amplitude and frequency. Or, in other words, for sounds of the same frequency the energy is proportional to the square of the amplitude, while for those of the same amplitude the energy is proportional to the square of the frequency.

Although the sensation of loudness is dependent upon intensity—in that an increase in intensity usually brings about an increase in the sensation of loudness—there is not a straightforward one to one relationship between them, otherwise, for instance, the sound of one piccolo could never be heard against the rest of the instruments in the orchestra.

Relative intensity or change in intensity is usually of greater practical importance than absolute intensity and is measured in logarithmic units called *decibels*. One bel corresponds to an increase in intensity of ten to one, while one *decibel* (one tenth of a bel) corresponds to an increase in intensity of the tenth root of this ratio, which is 1·26 to 1. Those who cannot follow this statement may easily verify that it is so by finding the continued product (see p. 250) of ten 1·26's (more accurately 1·2589 . . .).

The intensity of the vibration which produces in the ear the maximum sensation of loudness that it can stand (termed the *threshold of feeling*) has, in that region of frequency to which the ear is most sensitive, about ten million, million times as much energy as that intensity corresponding to the faintest sound that can be detected (termed the *threshold of hearing*). The only practicable way to cope with such a wide range of values is to forsake linear (common difference) units for logarithmic (common ratio) units.

Whenever two simple vibrations in the air have slightly different frequencies they are constantly falling in and out of step. The result is that the physical intensity of the combined vibration in the air is constantly decreasing and then increasing at a rate equal to the difference of the frequencies of the two vibrations. This ebb and flow or pulsation in the intensity of the combined vibration is known as a *physical beat*. When displayed on the screen of an oscilloscope physical beats appear in the form shown in the lower diagram on page 26 of Chapter 3, *i.e.* as periodic constrictions and expansions, and represent throbs in the intensity of the combined vibration. This figure indicates two such throbs or beats.

If these two simple vibrations are replaced by complex periodic vibrations then things become much more complicated. For if—as, in effect, does the ear—we regard each of the complex vibrations as consisting of a number of partial vibrations, it becomes obvious that, in addition to those previously occurring between their fundamental partial vibrations, beats can now occur simultaneously every pair of upper partial vibrations. For instance, in the time taken for one beat between the fundamentals of such a pair of vibrations, corresponding to a mistuned unison, there will now be two beats between their second partials, three beats between their third partials, etc. For other mistuned consonant intervals the situation becomes even more complex, and the complete set of beat rates can only be calculated if the fundamental frequency of each vibration is known. Though, of course, in all cases where the beat rate of any pair of partials is known, the corresponding pair of partials one octave above must beat at twice this rate—and so on.

Since the periods of each of the harmonic upper partial vibrations of a complex periodic vibration recur exactly in the periodic time of its fundamental vibration, it follows that no beating is possible between any *harmonic* upper partial vibration and its *fundamental* vibration.

12

INTERVAL

Interval—standardisation by string-length ratios—frequency ratios and string-length ratios—measurement of intervals in logarithmic units—standard form—melodic intervals—harmonic intervals—consonant intervals—interval definition—order of consonance—tetrachord—diatonic—diatonic intervals—major diatonic scale—chromatic intervals—step—major tone—minor tone—diatonic semitone—whole tone—mutable interval—comma of Didymus—comma of Pythagoras—chromatic semitones—harmonic seventh—inversion of intervals—complement—first and second inversions of triads—triads—consonant triads—dissonant triads—major and minor triads.

Unless great care is taken to distinguish them, the name or 'label' given initially to something purely for convenience becomes, in the process of time, confused with or even mistaken for the thing itself. The fifth, for instance, is presumably so-called because it occurs in Western music between most notes which are five degrees apart, though otherwise the interval itself has nothing to do with 'five' or 'one-fifth'—even when considerations are restricted to the purely physical or mathematical. The same interval is recognised and used almost everywhere in the world under various names.

A musical *interval* is a relation between two notes recognised solely through the ears and brain, *i.e. is an effect on the aural perceptions.* This being so, it follows that, if music is to be allowed to give the greatest satisfaction, then the choice of intervals most appropriate to the musical occasion must be left entirely to the judgement exercised by the sensitive ear and musical instinct of the skilled musician—and not be predetermined for him by the theorists, from purely physical or mathematical considerations.

If any two notes are sounded the musical ear compares them, in some subtle way, from the effects produced in it by their corresponding vibrations. Further, any pairs of notes—whatever their pitches—of which the corresponding frequency ratios are the same, are accepted by the ear of the musician as having a similar musical relation to one another, *i.e.* are recognised as more or less the same musical interval. On this fact depends the existence of what are known as musical scales—scales which are the same at any pitch and repeat at the octave. This correlation between frequency ratio and musical interval may be very exact in a consonant (harmonic) interval, though it depends also on the varying sensitivity of the listeners' ears, etc. It is, for instance, possible for the trained ear of the musician to perceive a consonant interval as being more true than the physical ratio of the vibrations producing it would suggest. This may happen especially when the interval is of short duration (such as one produced by notes from plucked or struck strings) or when the definition of the notes forming the interval is somehwat vague (such as are produced by an instrument having a soft tone, *e.g.* flute, etc. or one which suffers from appreciable inharmonicity.

When consonance is not involved the correlation between pitch and frequency is generally not so close. (See Chapter 13, *pitch* and last few paragraphs.)

The notes forming any interval *chosen by the ear* can be found experimentally by sounding appropriate lengths of a divided string—such as a monochord—and the *ratio of the string-lengths* so obtained then used to *standardise the interval physically*—and vice versa. But see also *inharmonicity, et seq.* in Chapter 11.

So far as the physics of musical sounds is concerned the vibration in the air is the end-product whatever the instrument used. It is the only consistent common factor shared by the ear of both artist and listener. There is no need whatever for either one of them to worry himself with such things as string or pipe lengths etc.—which are, in any case, variously affected by physical factors outside their control, such as temperature, humidity, tension. Hence, intervals are best standardised physically in terms of the ratios of the frequencies of the vibrations that they produce in the air, and, unless otherwise stated, all ratios used in this book are frequency ratios.

The frequency ratio corresponding to any given interval is the reciprocal of the corresponding string-length ratio. For instance, a frequency ratio of $\frac{5}{4}$ would correspond to a string-length ratio of $\frac{4}{5}$—and vice versa. See Experiment V in Chapter 18.

Though frequency may be used to specify, define, fix, or standardise intervals physically, they are not themselves intervals, neither are such frequency ratios directly proportional to the width of the intervals they are used to define, and consequently they do not give a true measure of these intervals. For if the latter were so, then it would be possible, in effect, to add (or subtract) intervals physically by simply adding (or subtracting) their corresponding frequency ratios—whereas, in fact, to *add* (or *subtract*) intervals physically it is necessary to *multiply* (or *divide*) their corresponding frequency ratios. While to *measure* intervals and represent their widths graphically in a linear manner it is necessary to *measure* their ratios, which means, in effect, to find their ratios in terms of *logarithmic units.* When so expressed, the physical sums and differences of intervals *can* be calculated by simple addition and subtraction of these units.

Because it is fundamental—at least to Western music—that all intervals be repeated at the octave, it is likewise more convenient, for the purpose of comparison, to quote all ratios relative to the 'octave' (or '$\frac{2}{1}$') in which they lie. This can be brought about by multiplying (or dividing) their ratios by 2, or by that power of 2 which makes them lie within the range of the ratios $\frac{1}{1}$ to $\frac{2}{1}$—*i.e.* putting them in a kind of *standard form* used in musical acoustics for quoting frequency ratios. Once this has been done it is a simple matter to convert the ratio to a decimal form, look up its corresponding value in cents from the Ratio Table, and hence get an accurate measure of the interval to which it refers. See also the beginning of Chapter 20.

A *melodic interval* is one formed when two notes are sounded in *succession*, whereas an *harmonic interval* is one formed by sounding two notes *simultaneously*. The accuracy with which it is possible for the ear to estimate an harmonic interval depends upon its definition (see Chapter 13—*aural definition*—and below).

In listening to notes related by octaves the human hearing faculty perceives a unique sense of identity; a sense not unlike that which it experiences with unisons, but which it finds impossible to associate, in general, with any other intervals. Thus, it is possible for confusion to occur, at least momentarily, between intervals of, say, two octaves and three octaves—especially when originating from different kinds of instruments. But no one confuses an interval of two fifths with an interval of three fifths—such as may be heard during the tuning of a violin.

Consonant (harmonic) intervals, such as the unison octave and fifth, for which the slightest mistuning produces perceptible beating between the component vibrations corresponding to the lower partial tones, are said to be *sharply defined*.

The consonant interval with the sharpest definition is the unison. For whereas in a perfet unison no beating is possible, with a mistuned unison beats will occur between *every* pair of partials of the same order (see Fig. A, p. 136) with the result that the strongest possible dissonance is experienced in the ear. Next in order comes the octave (see also *perfect consonances*, Chapter 13).

The larger the numbers which form the ratio of such an interval, the higher—and generally the weaker—will be the pairs of partial tones formed between the two notes that will beat in the ear as imperfect unisons when the interval is mistuned; the more will this beating be masked by the presence of the lower partial tones of both notes and by possible beating between them—to say nothing of difference tones—and consequently *the less sharp will be the aural definition of the interval.*

Thus the *order of consonance* of intervals—say, within the octave—will be given, in general by those corresponding to the following (frequency) ratios:—$\frac{1}{1}$, $\frac{2}{1}$, $\frac{3}{2}$, $\frac{4}{3}$, $\frac{5}{3}$, $\frac{5}{4}$, $\frac{6}{5}$, and $\frac{8}{5}$, *i.e.* the unison, octave, fifth, fourth, major sixth, major third, minor third, and minor sixth.

When widened by an octave, the fifth becomes a twelfth—a perfect consonance of ratio $\frac{3}{1}$; the major third becomes a major tenth—a more consonant interval than the major third and of ratio $\frac{5}{2}$. When widened by two octaves the major third becomes a major seventeenth—a perfect consonance of ratio $\frac{5}{1}$. All the remaining consonant intervals lying within the octave become less consonant by being widened by an octave. The fourth, for instance, becomes an eleventh, ratio $\frac{8}{3}$.

The *fact* that the octave is more sharply defined than the fifth *is decided solely by the ear. This fact cannot be over-emphasized.* That the arithmetic ratio used to define the octave *physically* is the simpler one is an interesting observation made *after the fact.* But such additional information means nothing to the ear and cannot in any way influence its judgement.

The basis of the Greek scale-system was the *tetrachord.* This was the name given to the four successive notes produced by four successive separate strings, the first and last of which were invariably tuned to give the interval of a fourth. Octaves were then regarded as made up from two tetrachords separated by a tone. Three possible ways of tuning the two inner strings were recognised, the resulting tetrachords being termed diatonic, chromatic and enharmonic. When so tuned as to give, in any order, the intervals of two tones and one semitone, the division was said to be *diatonic.*

The intervals found in a diatonic scale (or mode), such as, for instance, the major diatonic scale (or Ionian mode) are known as *diatonic intervals.*

In a *major scale* all the diatonic intervals measured upwards from the tonic are either major or perfect—hence its name—although all the other diatonic intervals, such as the minor third, sixth and seventh, and the diatonic semitone and tritone, can be found elsewhere in the scale between pairs of notes which do include the tonic.

Those intervals which are not found in a diatonic scale are called *chromatic intervals* in contradistinction to 'diatonic'.

The intervals between adjacent degrees of a scale are called *steps.*

The fundamental steps of the diatonic scale are the major tone, the minor tone and the semitone. All other diatonic intervals can be expressed in terms of the sum of two or more of these diatonic steps (see Chapter 19 and Figs. C and D).

The amount by which a fifth exceeds a fourth *i.e.* the interval which corresponds to a frequency ratio of $\frac{9}{8}$ (203·9 cents) is called a *major* (or *greater*) *tone.*

A *minor* (or *lesser*) *tone* is the amount by which a major sixth exceeds a fifth *i.e.*, that interval which corresponds to a frequency ratio of $\frac{10}{9}$ (182·4 cents).

The amount by which a fourth exceeds a major third *i.e.*, the interval corresponding to a frequency ratio of $\frac{16}{15}$ (111·7 cents) is termed a *diatonic semitone*.

The interval between adjacent degrees of a diatonic scale—excepting those between which a diatonic semitone is formed—is, in general, termed a *whole tone*. Thus a whole tone may be as much as a major tone or no more than a minor tone, depending upon the requirements imposed by the musical circumstances, *i.e.*, is generally a *mutable interval*. In all the practicable regular temperaments (*e.g.* equal and mean-tone) all whole tones are of fixed and equal value.

1 whole tone of equal temperament = 200 cents (approx. ratio $\frac{449}{400}$).

1 whole tone of mean-tone temperament = 193·15 cents (ratio $\frac{\sqrt{5}}{2}$) which is the (geometric) mean of a major tone (ratio $\frac{9}{8}$ or 203·9 cents) and a minor tone (ratio $\frac{10}{9}$ or 182·4 cents) which is the same thing as the (geometric) mean of a major third (ratio $\frac{5}{4}$ or 386·3 cents).

The amount by which four fifths exceed the sum of two octaves and a major third is called a *comma* (*of Didymus*), or sometimes a *syntonic comma*. This is also the amount by which a major tone (T or 'ratio $\frac{9}{8}$') exceeds a minor tone (t or 'ratio $\frac{10}{9}$'). It corresponds to a ratio of $\frac{81}{80}$ which equals 21·506 [21·506290] cents or 5·395 savarts.

In musical performance, this comma should not be thought of as the problem it presents to the tuner of keyboard instruments, but as a somewhat elastic interval depending, as it must, on the ear's ability to estimate intervals under varying musical circumstances. Further, if the musical instinct of an artist for melody or for concord compels him to make a small adjustment of this kind in the width of his intervals, then any subsequent readjustment felt necessary can easily be absorbed, without offending the ear, in a discord, which, because of its dissonance, is lacking in definition. Such an adjustment on the lower part of a cello string could require a change of about half an inch in its sounding length.

The amount by which twelve fifths exceed seven octaves is

called a *comma of Pythagoras* and has a frequency ratio of $\frac{531441}{524288}$ In logarithmic units this equals 23·5 cents or 5·9 savarts.

To change a major third (or major sixth) to a minor third (or minor sixth) it is necessary to reduce the interval by an amount equal to

$$\begin{aligned} M3 - m3 &= T + t - (T + s) && \text{or ratio } \tfrac{5}{4} \div \tfrac{6}{5} \\ &= T + t - T - s && \textit{i.e. } \tfrac{5}{4} \times \tfrac{5}{6} \\ &= t - s && \text{or } \tfrac{25}{24} \\ &= 182{\cdot}404 - 111{\cdot}731 \\ &= 70{\cdot}673 \text{ cents.} \end{aligned}$$

Whenever this is done to a diatonic interval the interval is said to be chromatically changed and the amount of change required is called a *chromatic semitone*. If, on the other hand, it is required to change a minor third (or minor sixth) to a major third (or major sixth) then the process is reversed, though there is one important exception. In any temperament which distinguishes between major and minor tones the interval $t + s$, ratio $\frac{32}{27}$, is possible (*e.g.* DF in just temperament). This interval is a comma short of a true minor third, $T + s$, and therefore needs a comma more than the chromatic semitone mentioned above, to make it into a true major third, *i.e.*, requires to be increased by an interval of ratio

$$\begin{aligned} T - s &= 203.910 - 111.731 \quad \text{or} \quad \tfrac{25}{24} \times \tfrac{81}{80} = \tfrac{135}{128} \\ &= 92{\cdot}179 \text{ cents} \end{aligned}$$

In order to distinguish these two chromatic semitones, the first, ratio $\frac{25}{24}$ is called the *smaller chromatic semitone* or *small half-tone*, and the second, ratio $\frac{135}{128}$ is called the *larger chromatic semitone* or the *small limma*. Definite ratios such as these can, of course, only be justly assigned to chromatic semitones derived harmonically—as above—or to specific tunings. When they occur melodically the intonation of chromatic semitones is given a degree of flexibility which is decided largely by the taste and feelings of the artist.

The *harmonic seventh* is the name given to the interval betwen the fourth partial tone (two octaves above the fundamental) and the seventh partial tone of a bowed string, and although Western music has not yet made use of it as a definite interval of its scale-system, it does come naturally in the notes of the trumpet, French horn, etc., and has been used occasionally in melodic passages by composers such as Vaughan Williams.

Its position in the scale is $\frac{14}{11}$ of a comma (about 27 cents) flatter than a true B♭ (ratio $\frac{16}{9}$) *i.e.*, an amount equal to twice the theoretical sharpening of the major third in equal temperament.

All the diatonic and chromatic intervals of our musical scale can be expressed in terms of the octave, the fifth and the major third and would in effect be so produced as part of the prevailing harmony in sixteenth-century polyphony. The interval of the harmonic seventh cannot however be so expressed; and this suggests that though it could occur, more or less by chance, in sixteenth-century polyphony, it would do so only as an unessential note, not forming part of the prevailing harmony.

The seventh partial tone of a note of an instrument, such as a violin, clarinet, etc.,—a pure tone—is said by some writers to be 'out of tune', 'dissonant' or even 'inharmonic'. Such statements are not only misleading but also quite untrue. (See Chapter 13, HEARING—*perfect consonances.*)

If either the lower of two notes forming an interval is raised an octave, or the higher is depressed an octave, then the interval is said to be *inverted*. It follows that any interval added to its inversion must equal an octave, or that the inversion of an interval must equal that interval by which an octave exceeds it.

Thus the inversion of a fifth,

= octave − fifth = a fourth

= (in terms of ratios) $\frac{2}{1} \div \frac{3}{2} = \frac{2}{1} \times \frac{2}{3} = \frac{4}{3}$

= (in terms of cents) 1200 − 702 = 498 cents

or the inversion of a major third,

= octave − major third = minor sixth

= (in terms of ratios) $\frac{2}{1} \div \frac{5}{4} = \frac{2}{1} \times \frac{4}{5} = \frac{8}{5}$

= (in terms of cents) 1200 − 386 = 814 cents

and so on.

Each one of such a pair of intervals is said to be the *complement* of the other.

Hence the inversion of an interval ratio, in standard form, can be obtained, by inverting it and multiplying it by 2, or, in terms of cents, by subtracting its value, in cents, from 1200.

In the case of a triad (see below) in its fundamental or closest position (*i.e.* with its notes so spaced as to form the smallest possible intervals) the combination obtained by raising the lowest note by an octave is called its *first inversion*. If the

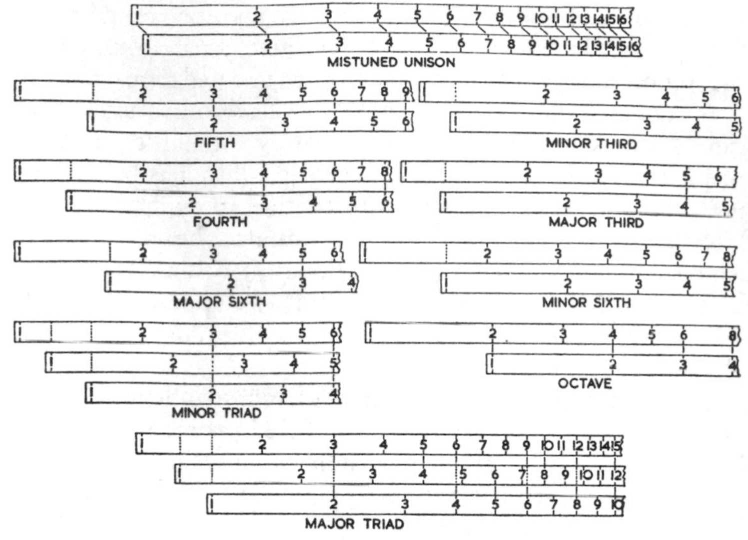

Fig. A

process is repeated with the lowest note of the first inversion then the resulting combination is called the *second inversion* of the triad in question. These inversions are shown graphically in Figure B below.

Though providing a convenient basis for classification, the use of the word 'inversion' should not be taken to imply that any one interval or chord owes its derivation to another.

Three notes sounded simultaneously—or the combination of two intervals having one note in common—make, between them, three intervals and are known collectively as a *triad*. In Fig. A, above, three sets of harmonic partial vibrations indicated by their numbers and corresponding to the notes of the major and minor triads are shown. In each case these partial vibrations are continued upwards until the first unison between three of them occurs.

If all the intervals so formed are consonant, then the combination is known as a *consonant triad*—if any one of them is dissonant the result is a *dissonant triad*.

The origin of these consonant triads can be traced back to the time when it first came to be considered as essential in all good polyphonic writing that—no matter how many lines there were—every line should make correct two-part harmony with every other line. One of the results of this was that certain seemingly fortuitous combinations of intervals, brought together by contrapuntal forces between the separately moving lines, occurred with sufficient frequency for them to become regarded as more or less separate entities without reference to their context. Thus there came into prominence the major and minor triads, what are now called their inversions, and several other diatonic chords.

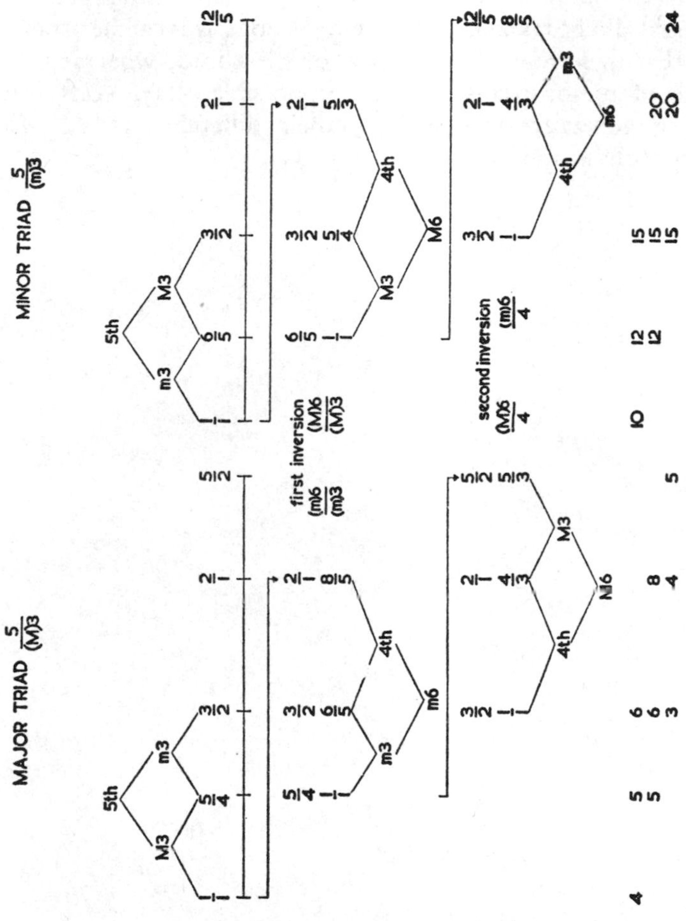

Fig. B

A *major triad* is one in which the greatest interval is a perfect fifth and in which the lower interval is a major third—the upper interval being then automatically a minor third. If this order is reversed and the minor third is made the lower of the two intervals included in the perfect fifth, then the combination is known as a *minor triad*.

The major and minor triads (and their inversions) are the only consonant (not necessarily concordant) triads available within the range of an octave.

In considering the consonance of triads, difference tones should also be taken into account. Major triads can be arranged so that these tones form a part of the chord, whereas in the case of minor triads this is not possible—their consonance being generally disturbed by their difference tones. (See Appendix 8, and pp. 18 and 19.)

13

HEARING AND LISTENING

Acoustics—sensations—listening and hearing—perceptions—phenomenon—partial tones—upper partial tones or overtones—simple (or pure) tones—Ohm's law—fundamental partial tone—harmonic upper partial tones or overtones—inharmonic upper partial tones or overtones—pitch—physical time—perceived time—rhythm—absolute pitch—standard musical pitch—physical beating—dissonance—unison—combination and difference tones—consonant intervals—concord—perfect consonances—aural definition of intervals—loudness—phon—masking—reverberation—cocktail party effect.

The science of hearing is called *acoustics* and as such is concerned with the study of the hearing-faculty, *i.e.*, with the partnership of the ear and brain. It starts with the vibration of the ear-drum and includes the physiological reactions in the ear that follow from it as well as the reactions of the nerve-endings of the nervous system of the ear and the responses they excite in the brain. Clearly, the structure and properties of the hearing-faculty must impose some natural limitations on the perception and comparison of sound waves. It is the object of scientific enquiry to ascertain these limitations and find out how people hear and what they hear when they listen to what they call a sound outside their ears. Music is part of the evidence which science has to take into account.

Impressions on the senses are called *sensations*. The nature of a sensation depends much more on the characteristics of the sense organ involved and the manner in which the brain co-operates with it, than on the nature of the external object or occurrence causing it.

Listening is the application of the mind's attention to sensations presented to it by the hearing faculty—a procedure which almost always involves some degree of selection of one kind or another. The process of *hearing* is dependent primarily on natural powers, whereas listening is more a matter of (musical) education and training in the exercise of these powers.

The mental images of external objects or occurrences which are produced from sensations are called *perceptions*. In this way particular kinds of musical sounds are perceived as associated with the instruments that produce them; while other kinds of sounds, containing different kinds of vibration, are perceived as vowels, etc.

A *phenomenon* is the psychologist's name for a product of sensory perception.

By concentrated selective listening it is possible for the ear to discover some of the component or *partial tones* in a single note, or tone—a fact first mentioned by Aristotle. In searching for the cause of this pehenomenon in the air-vibration it was found that, in general, only when a complex vibration was produced were component tones detectable by the ear, and that, further, any partial tones discovered by the ear were found to correspond with those partial or component (simple) vibrations into which it was possible to resolve the complex vibration, mathematically, by means of a Fourier's analysis, or physically, by, for instance, obtaining the natural harmonics of the open string of a violin. No *upper partial tones* or *overtones* could be discovered in the effect on the hearing faculty of a simple vibration of moderate intensity[1]—which for this reason was called a *simple* (or *pure*) *tone*. Thus *all partial tones are pure auditory senstations consisting of simple tones*, and *all complex periodic vibrations are, in the initial stages of our hearing process, resolved into a series of simple vibrations by the mechanism of the ear and then converted into a corresponding series of partial tones.* This statement—which establishes a relationship between a vibration listened to and the corresponding musical note heard—is known as *Ohm's law* of acoustics.

That this is so would be far more obvious but for the fact that

[1] Because of the non-linear response in our hearing faculty, any simple vibration will, if allowed to develop sufficient intensity, produce in the ear its own (subjective) harmonic upper partial tones.

these sensations called partial tones are not normally the subject of our conscious perception. The natural thing for the combination of ear and brain to do with any single complex periodic vibration, after having converted it into a series of partial tones, is to blend these sensations, by aural perception, so as to form one consistent and coherent whole, which it then hears, in the light of experience, as a complex note, having more or less the same pitch as that normally attributed to the *fundamental partial tone* and of characteristic quality and loudness depending on the number and relative strength of its various partial tones. This means that in order to become aware of the *individual* partial tones in a single complex tone, it is first necessary to learn to undo the naturally accumulated experience acquired during a lifetime spent in blending and interpreting them as a whole. Nevertheless, if there is available a simple tone of the pitch of the particular partial tone whose detection is desired, it will be found comparatively easy to transfer the attention from this simple tone to the particular partial of the complex tone, since the two partial tones will be identical in pitch and tone-colour. However, in general, it is only found possible to attend to one such partial tone at a time—the remaining partials being heard as a blended whole.

Partial tones are numbered from the fundamental tone upwards, *i.e.*, the lowest possible in pitch being termed the first—as is the case in frequency with their corresponding partial vibrations—and so on. Overtones—otherwise identical with partial tones—do not include the fundamental tone and may thus be said to consist of upper partial tones. However, as all partial tones are of the same nature, it is purely artificial to split them into a fundamental tone plus overtones—especially since the term fundamental tone implies the existence of overtones.

Upper partial tones, or overtones, may be either harmonic or inharmonic. *Harmonic* upper partial tones or overtones have pitches which correspond with partial vibrations whose periods form, together with that of their fundamental, an harmonic series. Those which do not so correspond, such as the overtones of bells, drums, struck bars, rods and strings, are called *inharmonic* upper partial tones or overtones.

Pitch is a perceptual effect in the brain. It is part of the response of the ear and brain to certain vibrations in the air—not something outside the ears that is listened to. Although it is true to say that the pitch of a note depends mainly on the repetition rate (frequency) of the time-displacement (vibration) pattern, it also depends to a greater or lesser extent, on the intensity (power) and harmonic content (complexity) of this vibration pattern—*especially when sounded on its own.* To what extent it does so depends also on the musical occasion and the musical sense of the listener.

A similar kind of variation from exact correlation as occurs between frequency and pitch is also found to exist between *physical time*—as measured in terms of the consistent physical oscillations associated with any kind of clock mechanism—and (real ?) *perceived time*. For instance, physical pulses of musical sounds of exactly equal duration are not, by any means, necessarily perceived as such, and will, in addition, vary somewhat from one listener to another according to individual training, degree of attention, musical context, etc. In this respect, the concept of *rhythm*—by which is meant *all* organization in time—is even more devious.

In this connection Dr. Fletcher[1] observed that two intense simple vibrations, the pitches of whose corresponding notes, when heard separately, were an octave apart, were found to be dissonant when heard together, because their frequency ratio was not exactly $\frac{2}{1}$; whereas, two other intense vibrations having the same fundamental frequency—one being simple and the other complex—when sounded separately, differed both in quality and pitch, but when sounded together, produced a single note, without dissonance, having a strengthened fundamental. The pitch in the latter case being found to lie between the pitches heard when the two 'notes' were sounded separately. These extremely interesting results in no way conflict with Helmholtz's theory of dissonance and investigation of the quality of musical sounds.

The strike-note of a bell cannot normally be accounted for in terms of physics and is a typical instance in which the sense of pitch is produced, not directly by frequency of vibration, but by aural perception of its sound.

The pitch associated with a complex vibration is not decided

[1] Journal of the Franklin Institute, 1935, ccxx, p. 426.

simply by the strength of its fundamental partial vibration—which may, in fact, be entirely absent—but is rather an effect produced in the brain that is characteristic of the complex vibration *considered as a whole.* See also *note* and *definition*, Chapter 14.

There is no such thing as *absolute pitch.* All so-called standards of pitch are physically based—relying as they do on some arbitrarily chosen reference frequency of vibration. The choice of the current *Standard Musical Pitch,* which is based on a frequency of 440 cycles per second for *a'*, was influenced largely by purely practical considerations—notably by the comparative ease with which such a frequency could be accurately generated (electrically) and then widely dispersed (by broadcasting)—and is the first such standard to meet with international approval (London 1939).

The need for some form of standard pitch would not have become strongly felt until it became common to use, in concert, instruments of a more or less fixed pitch—most wind instruments, for example. But even so, such standards would only operate locally at first, and have in fact varied considerably even down to relatively modern times.

If two simple periodic vibrations—whose corresponding notes form an interval somewhat less than a minor third in the treble stave—are generated simultaneously, then the *physical beating* that occurs between them causes an unpleasant sensation of roughness or *dissonance* in the ear. Reducing continuously this interval, though reducing the rate of physical beating, is found to increase the roughness, a maximum being reached when the rate of beating is about thirty per second *i.e.*, when the interval is about a semitone—any further reduction of the interval being accompanied by a fairly rapid reduction in the roughness experienced as the *unison* (beatless condition) is approached. When finally a unison is reached the two tones sound together at the same pitch as one—no beating then being possible as (ideally) the frequencies of the vibrations causing them are identical.

In the extreme treble the interval of maximum dissonance between two simple tones is considerably less than a semitone, while in the extreme bass it is about a major third.

Thus *dissonance* depends in a rather complex manner—

decided by the construction and characteristics of the ear and brain—on *rate* of beating. If beats of sufficiently high frequency occur no sensation is felt, so that musically they have no existence. Very slow beats are perceived simply as a slow waxing and waning of the tone—no dissonance is felt, and their effect is not necessarily unpleasant (*e.g.*, voix celeste). They can be very useful. The most accurate physical tuning by ear is only made possible by the detection of these slow beats.

If any two simple periodic vibrations are generated simultaneously and with sufficient intensity, then besides the corresponding simple tones, the ear, due to its asymmetrical construction, may hear *additional* so-called *combination tones*[1]—the most important of these being termed the (*first*) *difference tone* because it corresponds in pitch with that simple tone which would originate from a vibration whose frequency equalled the difference in frequency between the two vibrations. See pp. 18, 19 and Appendix 8.

Beats heard from mistuned octaves and fifths—formed from simple tones—owe their existence in the ear to these difference tones.

The term 'difference tone' is used throughout this book to mean *first* difference tone.

Further, if the two simple periodic vibrations are now replaced by complex ones—*e.g.* by 'notes' from musical instruments—then, in general, beating will occur, not only between the lowest partial tones or fundamentals of each note—but also between the many upper partial tones. Moreover, the rate of beating between any adjacent pair of partial tones will now depend, not only on the interval between their two fundamentals, but also on the order of the periods of their corresponding partial vibrations in the harmonic series. For example, taking the case of a mistuned unison between fundamentals—*i.e.* between the two notes, as indicated diagrammatically in Fig. A—in the time taken for one beat to occur between fundamentals, two would occur between second partial tones, three between third partial tones, and so on.

Combination tones—especially difference tones—will, of

[1] These subjective tones are in addition to those harmonic upper partial tones already mentioned in the previous footnote, and can, in special circumstances, be shown to have actual physical counterparts in the air's vibration.

course, also be produced in the ear by such a pair of complex periodic vibrations, and although naturally more involved and thus less easily perceived *separately* as simple tones do nevertheless have some effect on consonances—more particularly so where triads are concerned.

All this was first tried out by Helmholtz. It was he who discovered that the dissonance produced by two notes sounded simultaneously was due to the sensation of roughness set up in the ear by beating between upper partial tones, and that for *certain exceptional intervals the roughness either passed through a minimum or was almost entirely absent.* Not only did these exceptional intervals coincide with those accepted by the art of music as consonant, but they also agreed so far as the order of consonance was concerned *i.e.* they were the *consonant intervals.* In this way Helmholtz gave a physiological meaning to consonance, and although, broadly speaking, it is treated by the art of music as the basis of concord, nevertheless *consonance should not be confused with concord—a purely musical term implying performance* (see p. 79, fn. 13). Good consonances will always sound in tune.

To return now to the last by one paragraph—the addition of upper partial tones to the fundamentals has two main effects.

The first is to extend indefinitely the range of intervals through which beating can occur, since if sufficient upper partial tones are present in each note there will almost always be some from the one note which will lie within beating distance of some from the other note.

The second effect is that both the amount of dissonance and its rate of increase with change of pitch, either side of a consonant interval, is greatly augmented. As a result of this even the slightest mistuning of those partials which should form unisons in a good consonance—as indicated in Fig. A—will now cause the ear to protest strongly against the dissonance so produced in it, and the interval is said to be given much stronger *definition* in the ear. There is something deeply satisfying in an accurate consonance between two rich tones. Acoustically speaking, both the number and the strength of the upper partial tones are of any note of great importance when considering the degree of dissonance of any interval it may make with another note: for instance, if the fifth partial tone of the lower note is at all powerful, then the interval of the major sixth may well sound less dissonant than the perfect fourth. If the notes sound at all loudly then further definition may be given to the interval from beating due to combination tones. Finally, since time is required by the ear to detect dissonance, it follows that the definition of an interval in the ear will be impaired if the notes forming it are of insufficient duration.

No beating, and therefore no dissonance is possible between any *harmonic* upper partial tone and its fundamental—in fact, any intervals between two notes whose fundamentals fall in the same harmonic series must be, in this sense, *perfect consonances*, *e.g.* the unison ($\frac{1}{1}$), the octave ($\frac{2}{1}$), the twelfth ($\frac{3}{1}$), the fifteenth or double octave ($\frac{4}{1}$), the seventeenth ($\frac{5}{1}$), etc. However, although all equally consonant, their sharpness of definition decreases as their width increases. For, while no beating is possible so long as they are in perfect tune, nevertheless, the number of partials which beat as imperfect unisons when each interval is slightly mistuned becomes less and less (and of higher and higher order) as their width is made greater. Thus, with the mistuned unison, beating occurs between every pair of partials of the same order; with the octave, every other pair (see Fig. A); with the twelfth, every third pair—and so on.

Dissonance is not a physical quantity, and cannot therefore be equated or measured in terms of other physical quantities. The sensation of beating, the roughness, is a physiological effect in the ear itself—an experience—and therefore only the ear can say that the octave, fifth and fourth are sharply defined, the thirds and sixth less so, and the outlines of discords vaguer still. As an example of the latter consider the tritone—a fourth, of ratio $\frac{45}{32}$—two major tones plus one minor tone, or a fifth minus a diatonic semitone. The ear has no physical means whatsoever of estimating the exact tuning of this interval. For, even assuming such high partials as the 45th and 32nd to be present in some strength in the notes, and to lie within the audio range, the ear would still be quite unable to detect any mistuning between them because of the masking (see later) and dissonance caused in it by the sounding and beating of the number of more powerful partials of lower order that would also be present. It might, of course, hit exactly on such an interval by a lucky chance, but this is another matter. Similar remarks would, of course, apply to its inversion, the imperfect fifth, ratio $\frac{64}{45}$, and to equally tempered intervals, etc. See pp. 74 and 77.

Thus the ability of the ear to estimate harmonic intervals depends first on its own limits of accuracy which are, in turn, limited by the duration of the notes, the definition of the intervals and the dissonances of the harmony within which it has to work. In addition it must be remembered that the smoothness of consonance varies with pitch—being less in the bass than in the treble. Finally, it should not be forgotten that the ear of a musician who is constantly playing an instrument with rigidly fixed intonation, such as an organ, may easily lose some of its sensitiveness for consonance—*i.e.* for absence of beating—because it is so often listening to music which is not exactly in tune.

Both richness of tone-colour and definition of single notes are considerably improved by the addition of upper partial tones, although if these extend over too wide a range and sound at all strongly then the note will contain its own element of dissonance and will sound rough or harsh because of beating between its upper partial tones, such as for instance, its 15th and 16th, which are a semitone apart. Such harshness might be much more of a problem if it were not for the fact that, in general, upper partial tones get weaker and weaker as their order in the series increases.

It is also possible, under certain conditions, for the loudness of a relatively simple tone to be increased many times by the addition of upper partial tones requiring very little extra expenditure of energy. This, for example, is why a reed stop on an organ sounds louder than a stopped diapason which uses the same amount of wind. For whereas the first is rich in upper partials, the second is very poor and has no even partials at all. The explanation lies in certain properties of the ear—some of which are mentioned in the following section.

Loudness is an aural sensation, which, when viewed from a purely physical standpoint, can be regarded as an impression of the apparent physical strength of audio-frequency vibrations attributed to them by the hearing faculty—depending on the intensity, frequency and harmonic content of the vibrations listened to. Since the degree of sensation felt varies considerably among individuals a standard of equivalent loudness, such as the phon, can only be arrived at by obtaining a kind of average sensation based on the judgment of a number of normal people. The loudness level in *phons*, produced by a simple vibration of any frequency, is defined as being equal, numerically, to that level of intensity of a simple 1000 cycle-per-second vibration (measured in decibels) which is judged by the ear to produce the same level of loudness. Decibels and phons are thus automatically made equal, numerically, at a frequency of 1000 cycles per second, though, since the sensitivity of the ear varies considerably both with frequency and intensity, they soon part company either side of this frequency. The ear is most sensitive at frequencies lying between about 3000 and 4000 cycles per second—corresponding with the top six notes of a piano. For simple vibrations either side of these values the ear becomes progressively less sensitive—particularly for those vibrations of low intensity and frequency associated with notes in and below the bass stave. That is why, for instance, although a good deal more energy is required to play the tuba than to play the piccolo the latter can nevertheless still hold its own as far as loudness is concerned. It also explains why the sounds of the large organ pipes in the bottom octave of the open diapason do not overwhelm the sounds of the smaller pipes in the upper octaves—which take far less wind *i.e.* far less energy, from the windchest than do the larger pipes.

Békésy[1] has pointed out that without this loss of sensitivity at low frequencies normal hearing would be seriously disturbed by the noise from vibrations generated in the body whenever the muscles or joints were moved or jarred—to say nothing of external vibrations of a similar nature.

In order to make its existence known to the hearing faculty a simple vibration corresponding to a note at the bottom of the bass stave needs nearly 10,000 times (40 decibels—see Chapter 11) the energy of the faintest vibration which can be detected at a pitch just above the treble stave. This makes the detection of low pitched difference tones extremely difficult.

When upper partials are brought in, the situation, as one might expect, becomes very much more complex. For instance, a tone perceived from a simple vibration of lower

[1] See Scientific American, Vol. 197, No. 2.

frequency tends to mask (see below) one perceived from a simple vibration of higher frequency. But, on the other hand, a tone corresponding to a simple vibration of higher frequency may lie in a region where the ear is more sensitive than it is at the lower region, and it is reasonably safe to say that if five harmonic partial vibrations were set in motion simultaneously, then the higher ones would contribute more than their 'fair' share to the loudness of the complete (complex) tone or note as heard.

The reduction in the ear's ability to detect one sound due to the simultaneous sounding of another is known as *masking*. The degree of masking experienced varies with both the frequency and the intensity of the masking vibration. In general, masking increases with intensity and decreases with frequency—but in neither case is the relationship a simple one.

It follows from all these observations, that a complex sound sensation is changed both in character and in loudness whenever the intensity of its corresponding complex vibration is increased or diminished—even when this is accomplished without distortion—and that the reduction of background noise to an absolute minimum is essential to good performance of music and faithful reproduction of musical sounds.

In the study of musical acoustics the musician naturally starts with pitch, loudness, quality, etc., because these are what he can hear. The physicist, on the other hand, starts with frequency, intensity, harmonic content, etc., because these are what he can measure. But such physical quantities cannot by their very nature be musical, and unless the physicist takes into account the properties of the hearing faculty his study can have no relevance in musical matters. It is, for instance, quite true that the ear is not the most accurate and reliable device for estimating the values of physical properties of vibrations such as intensity, frequency, etc.—but to say, or imply, as many people have, that what the ear hears must therefore be 'incorrect', is quite absurd and meaningless. The making of physical measurements is not the main function of our hearing faculty and must not be assumed to be so.

Some of the important properties and processes involved in the link between the physical and the musical are shown diagrammatically on p. 150.

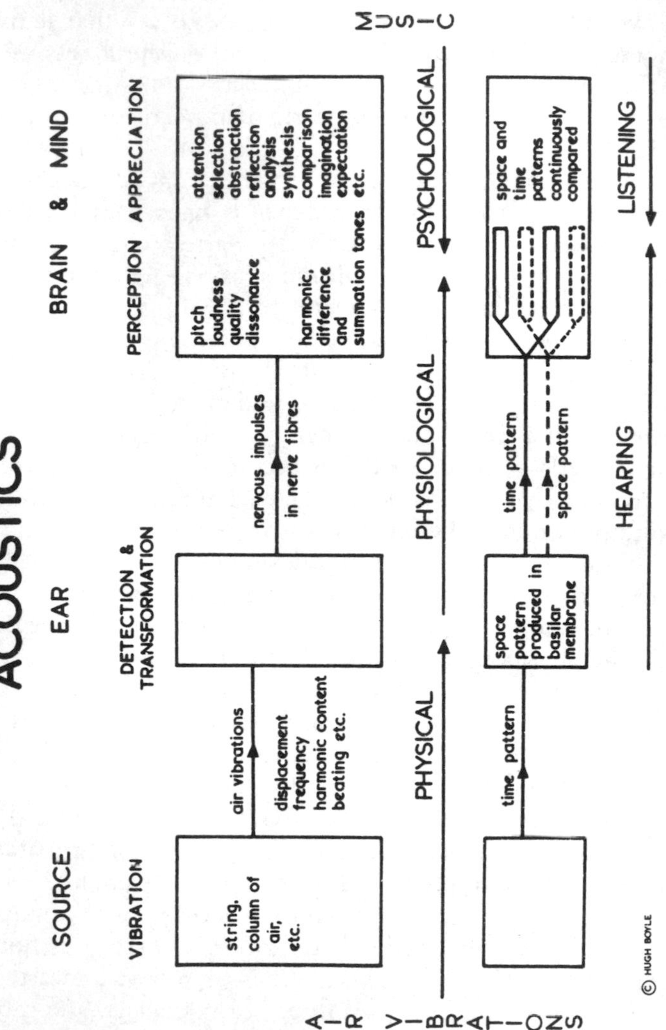
ACOUSTICS
SOURCE
EAR
BRAIN & MIND
VIBRATION
DETECTION & TRANSFORMATION
PERCEPTION
APPRECIATION
string, column of air, etc.
air vibrations
displacement frequency harmonic content beating etc.
nervous impulses
in nerve fibres
pitch loudness quality dissonance
harmonic, difference and summation tones
attention selection abstraction reflection analysis synthesis comparison imagination expectation etc.
AIR VIBRATIONS
PHYSICAL
PHYSIOLOGICAL
PSYCHOLOGICAL
MUSIC
time pattern
space pattern produced in basilar membrane
time pattern
space pattern
space and time patterns continuously compared
HEARING
LISTENING
© HUGH BOYLE

But this is a most difficult subject and one on which a great deal of work has yet to be done. The main object of the diagram is to show how distant is the connection between vibrations and music, how varied and complex the links between them, and how unreasonable it is to make deductions about sound sensations and perceptions from observations and experiments made on air vibrations *alone*.

Some additional facts worth bearing in mind which affect our hearing and listening are given briefly below.

It is impossible for the ear to attribute a definite pitch to any periodic vibration unless it continues for at least $\frac{1}{100}$ of a second. Under exceptional circumstances this might well have to exceed $\frac{1}{10}$ of a second—the exact time required depending on the frequency, the ear of the listener, etc. In addition, due to the ear's 'persistence of audition' the sound will continue to be heard for about the same length of time after the vibration has ceased. Thus the ear cannot possibly produce perceptible changes in the brain corresponding to each single vibration or subtle instantaneous variation that might occur in an otherwise purely periodic vibration—such as small fluctuations in frequency and waveform—but must attribute to such a vibration a kind of blended sensation continuously averaged over the number of vibrations executed in some small fraction of the second that has just elapsed.

Musically speaking the term 'steady state' is a relative one. First of all, there are limits to the degree of accuracy to which this state can be maintained. Such limits are, in any particular musical circumstances, predetermined for us by the smallest deviations from steady-state conditions that our ears are able to detect—the result being a kind of 'average steady state'. Secondly, our ears, and in fact all our senses, respond much more readily to *changes* in stimulus, or put in another way, very quickly lose their sensitivity to steady state conditions. Thus it is not perhaps so surprising to find that with those instruments on which the skilled musician is allowed to both form and continuously control his notes, he can, by introducing small yet perceptible deviations of a suitable nature, endow them with 'life', 'warmth' etc., while at the same time preserving their more or less 'average steady state'.

The situation when dealing with damped vibrations is,

needless to say, a much more difficult and complex one—whether considered from the physical or from the physiological standpoint. See also pp. 7 and 8, and Chapter 14.

The ear's response is affected by the (involuntary) change in the degree of tension (or relaxation) in the muscles of the middle ear brought about by alterations in the general level of intensity of vibration outside it.

The original vibration is often considerably modified by the superimposing of echoes—such as the multiple diffusion of echoes which occurs in any enclosed space—particularly if the space is a large one with good sound reflecting surfaces. Those which persist after the original vibration has ceased are known as *reverberation*. However, with sharp transient sounds, the effects of this condition may be somewhat offset by our automatic echo-suppressing mechanism—which is effective for a fraction of a second after the original sound has ceased—and, more generally, by moving the head so as to bring in the directional properties of our hearing.

Even without training we all possess some ability to select from a number of sounds those which we want to hear (sometimes called the '*cocktail party effect*'). Exactly how this is brought about has not yet been discovered. But there does seem to be some form of 'feedback' mechanism from the brain to the ear that may well be employed whenever we wish to exercise a degree of selection from the many sound sensations presented simultaneously to the brain by the ear.

The ear's power to separate notes which are close together in pitch is limited—the extent of this limitation varying considerably with the circumstances and the individual.

For example, until there is a difference in pitch of at least a comma between two notes when sounded *separately*, the ear finds it impossible to separate them when sounded *simultaneously*. It hears only one note accompanied by beating—the pitch of this note lying between those pitches heard when the notes are sounded separately. *It does not hear two separate notes*. Only the presence of beating tells the ear that the unison is imperfect. But the whole of this matter is dealt with in much greater detail in Chapter 3.

Among other things, this property of our hearing makes clear how impossible it is to obtain any comparison between two different kinds of fixed tunings by sounding their corresponding notes *simultaneously*.

It also helps to explain why, for instance, the first violins of an orchestra, when playing in unison, are heard to have a definite pitch despite the fact that, due to different rates of vibrato, etc., the various physical vibrations produced simultaneously cover an appreciable range of frequencies.

Another important point to realise is that almost any form of distortion can be, and frequently is, accommodated by the ear. In fact, unless the distortion is of an exceptional nature, the ear can rapidly become so accustomed to it as to lose all consciousness of its existence. That this is so should appear obvious to anyone who really listens to the generally accepted standards of reproduction of music and speech. It is, of course, equally obvious that the only way to counter this tendency is to make a habit of listening intently to live performances as often as possible.

Much excellent additional material on this subject—as, indeed, on many other associated subjects, such as the effects of non-linearity and of room acoustics on our hearing processes—may be found in Arthur H. Benade's *Fundamentals of Musical Acoustics*, published by the O.U.P. See Ref. FR2.

Appendices 1, 2 and 8, etc., also apply.

Finally, in all experiments in hearing it is well to remember Lord Rayleigh's remark in footnote 1 on page 15 of his *Theory of Sound:*—

> Most probably the power of attending to the important and ignoring the unimportant part of our sensations is to a great extent inherited—to how great an extent we shall perhaps never know.

14

NOTE

Note—tone—quality (goodness)—*definition—quality* (tone-colour or timbre)—*starting transient—attack—notes of the harmonic series* (*of acoustics*)—*chord of nature—unessential notes—melodic intonation—mutable notes.*

Music consists essentially of ordered progressions of intervals, melodic or harmonic, and to form intervals notes are used. In music, notes have no existence independently of intervals. In spite of this fact many people, without perhaps consciously realising it, make the assumption that music consists of notes, and that notes exist only at those certain arbitrarily fixed pitches which they have become accustomed to hear from keyboard instruments tuned in accordance with the present day physically standardised system.

Briefly, a *note* is the response of the hearing faculty to a complex vibration in the air caused by a musical instrument—including, of course, the most fundamental of these, the human voice. But the vibration itself cannot be the note, for it is impossible to perceive such a vibration *as a vibration*. That part of the brain which deals with our hearing can only interpret the language of discrete nervous impulses of the kind sent to it by the ear—as a result of air vibrations outside it of appropriate frequency and intensity. Vibrations in the air are of significance to the musician only in as much as they affect his aural perceptions. His sensitive ear, and the practical expression of its properties which is embodied in music itself, are all that matter to him.

Without human hearing there could be no notes, intervals,

scales or temperaments, though bodies would still vibrate in accordance with the laws of dynamics and produce corresponding waves in the air.

In more detail—a note is a phenomenon, or mental image, produced by aural perception from the sensations excited in the nervous mechanism of the ear called partial tones (see HEARING) which the ear and brain blend into a whole 'something' having a definite pitch, loudness and quality (goodness and tone-colour). In this sense notes have no existence outside the head.

'Tone' is also used for this purpose, particularly in science. This is sometimes enlarged to 'complex tone' to make a contrast with 'simple (or pure) tone'—the two terms then being analogous with 'complex vibration' and 'simple or pendular vibration'.

By extension, a musical instrument may be said to produce a 'note' or tone whenever it causes a more or less periodic vibration of such a frequency as to excite a response in the sensory apparatus of the ear. In this sense the 'note' in that part of its journey from the instrument through the air to the ear, can be thought of as taking the form of a complex periodic vibration in the air, though strictly speaking, no simple single term has yet been invented to describe such a vibration.

The sole judge of the quality (goodness) of the note of an instrument is the musical ear. The partial (simple) vibrations from an instrument whose 'notes' are maintained—*e.g.* by blowing or bowing—are harmonic, *i.e.*, their periods do not depart by any appreciable extent from the harmonic series (or frequencies from the arithmetic series 1, 2, 3, 4, etc.) with the result that the corresponding partial tones heard blend so smoothly that the result is perceived as a single note of characteristic quality and definite pitch. Such instruments produce true periodic vibrations in the air and are said to have a sweet tone. On the other hand, the sound of untuned church bells, or of drums, etc., contains inharmonic partial vibrations which consequently are perceived as jangling or rough sounds of a more or less indefinite pitch. All instruments that are plucked or struck, *i.e.*, allowed to vibrate freely, produce damped vibrations (see MUSICAL VIBRATIONS—*free vibrations*) in the air and fall into this category. For example, the note of a piano is not perceived as having such a definite pitch as is that—say—of an oboe.

Some instruments produce notes which are lacking in harmonic upper partial tones and consist of nearly pure tones, slightly reinforced perhaps with one or two octave harmonics. Such a note is sounded by a stopped organ pipe or by a flute blown softly. A note of this nature may be said to lack *definition* in the sense that if used in accompaniment to—say—vocal music, it does not assert itself aggressively against loss or gain of pitch by the singers.

The ear is never quite sure of the pitches of simple tones—such as those heard from tuning forks or isolated partial vibrations—and frequently misjudges the octaves in which they lie and the simple (physical) interval between them. Such tones are obviously unsuitable for any practical experiments in intonation.

On the other hand, the effect on the ear of the several harmonic partial tones of a good musical note is to intensify the unity and consistency of the note as a whole, giving it *definition* and individuality, and thereby to enable such notes to be perceived more or less independently of each other when sounded simultaneously.

The vibrations corresponding to such sets of partial tones may be regarded as exciting a tidy, regular pattern on the basilar membrane (a part of the inner ear), which, when transferred to the brain, is blended by aural perception into one musical whole and perceived as a musical note of a definite pitch. Whereas sets of vibrations corresponding to *inharmonic* partial tones may be thought of as producing an untidy, irregular pattern which is perceived as a jangle. *Herein lies the sole scientific significance of the harmonic series in matters of music.*

Well defined intervals can only be perceived from well defined, clean, steady notes.

Much of our scientific knowledge of hearing is due to the work of Dr. Harvey Fletcher and his associates of the Bell Telephone Laboratories. In one experiment Dr. Fletcher set up ten independent generators of simple vibrations having frequencies of 100, 200, 300, etc., up to 1000 cycles per second. When all ten vibrations were set in motion simultaneously with equal intensity he found that the ear insisted on hearing them as a single musical note of definite pitch corresponding to that associated with 100 cycles per second. He then tried the effect of eliminating some of the component vibrations and found that as long as there remained at least three consecutive component vibrations the ear's reaction, so far as pitch was concerned, was likely to remain substantially the same. Following this Dr. Fletcher tried a second experiment with a set of simple vibrations having frequencies of 100, 300, 500, 700 and 900 cycles per second. This time, though he could quite distinctly hear a (subjective) tone corresponding to 200 cycles per second, (the common difference frequency of this series) he found that the combination was perceived by the ear as a noise of indefinite pitch. These two experiments tend to confirm the fact stated above about harmonic partial tones. For they show that a common *difference* in frequency between consecutive components of a set of simple vibrations is not by itself sufficient to produce in the ear a single note of definite pitch. It would seem that each component frequency needs also to be an exact *multiple* of the common difference frequency. In other words, it appears desirable that the component frequencies should fall in the ratio of the arithmetic series 1, 2, 3, etc., or what amounts to the same thing, that the component periods fall in the ratio of the harmonic series, 1, $\frac{1}{2}$, $\frac{1}{3}$ etc.

That written above applies more particularly to periodic vibrations. But this is exceedingly dangerous ground on which to generalise—especially where damped vibrations are concerned. A tuned bell, for instance, spans three octaves with its first ten partials (two more than are found in the same span of the harmonic series) and is perceived as having a pitch an octave above its first partial—which is an octave below its fifth partial tone. Furthermore, in such a case the degree of damping varies continuously, not only from one partial to another at any given instant, but also from one instant to another for any given partial. The partial vibrations of a bell, unlike those of a blown pipe or bowed string, are more or less independent of each other and do not naturally form an harmonic series—though it is the bell tuner's job to change this and make sure that at least some of the lower ones do so.

For some fascinating accounts of experiments relevant to this matter see Chapter 5 and pp. 318 and 319, etc., of Ref. FR2.

The difference in *quality* (variety in tone-colour or timbre) between notes—say, of the same pitch but originating from different types of instruments—depends solely on the differences in the complexity (harmonic content) of the vibrations causing them. Some comparison of these complexities may be made by examining the associated displacement curves of the vibrations in question with an oscilloscope, though this is only possible with single notes resulting from periodic (maintained) vibrations. Using the mathematical tool of harmonic analysis, it is possible, in the above mentioned case only, to analyse the associated displacement curves so obtained into their component or partial vibrations—if one is prepared to take enough time and trouble. But, as stated in the last paragraph, the ear is not so limited, for if a number of notes are played or sung simultaneously, it is able to analyse the effects of the very complex periodic vibration into separate notes instantly with very little effort. The ear does this continually with little effort whenever music is recognised and appreciated.

The tone-colour of the notes perceived from a musical instrument is often referred to as being bright, dull, rich, poor, harsh, smooth, full, thin, round, hollow, etc. Such a description applies to the more or less uniformly sustained portion of the musical sound and is generally known as its quality. Thus the ear, when presented with a combination of, say, trumpet and flute, is able to distinguish the notes of the separate instruments, and, whatever their duration, to maintain that distinction simply by the contrast it continuously perceives in their quality. This difference in *quality* (tone-colour or timbre) between notes of the same pitch originating from different instruments depends solely on differences in the complexity (harmonic content) of the vibrations from which they originate. Some kind of comparision of these complexities can be made by examining the associated displacement curves of the vibrations in question with an oscilloscope—though this is only possible with single notes sounded on their own and generated in the ear from periodic (*i.e.* maintained) vibrations. By using the mathematical tool of harmonic analysis, it is then possible, in the above case only, to analyse the associated displacement curves so obtained into their component or partial vibrations—if one is prepared to take enough time and trouble over it.

But, as already stated, the ear is not limited in this way. For if a number of notes are played or sung simultaneously, it is able, with very little conscious effort and in a very short time, to analyse the effects on it of the extremely complex resultant vibration by arranging all the partial tones that it can discover into harmonically related groups which it then perceives as the separate complete notes—a masterpiece in the art of unscrambling. *This only the ear can do;* and it does it continually with an ease and certainty that completely conceals its achievement whenever music is recognised and appreciated—in fact the whole possibility of music depends on it. In contrast, the eye, when confronted with the same physical information—even when it is conveniently displayed for it on the screen of a cathode ray tube—is quite unable to cope. It cannot perceive any simple relationships of any sort and very quickly loses all interest in the matter.

A certain amount of change in quality (and loudness) with pitch is of course a typically recognisable characteristic of any normal instrument—the human voice for instance—though such a change is usually distributed fairly smoothly over the instrument's range. This characteristic can often be of help not only in distinguishing one class of instrument from another, but also, to a lesser extent, in detecting the more delicate distinction between two instruments of the same class.

There is, however, no doubt that in many cases the ear is unable to tell one instrument from another with sufficient certainty when judged by quality alone. Thus in certain circumstances it is possible to confuse the sounds of a French horn with that of a singing voice, or of an oboe with that of a violin, etc. Nevertheless, some degree of discrimination may still be possible if more attention can be paid to the characteristic noises—such as the scraping of a bow, the hissing of air, etc.—which often accompany sustained notes but which tend to be rejected or only heard subconsciously by the listening musical ear.

However, in spite of the previous remarks—which apply to tone quality only—the most important distinguishing characteristic of the sound of the note perceived from a musical instrument is due to the way in which its vibration behaves *before* it achieves a steady state—*i.e.* the nature of its build-up or *starting transient*. This is, in most instruments, under the direct control of the skilled musician and is known as his *attack*. The nature of these transients or attacks seems to be able to influence quite considerably the manner in which the *whole* of the vibration is processed and hence perceived by the brain's aural system.

With regard to the previous paragraph, it is important to realise—for instance—that in the Baroque period, when so much music was of a contrapuntal nature, the notes of most instruments were considerably less powerful than they are today, and that this (reduced) power was mostly concentrated in the higher harmonics. Further, these notes also had, relatively speaking, a stronger and keener attack. This sharper attack helped both to command the ear's attention and to separate the individual parts, while the lower volume practically eliminated any masking (*q.v.*) between them. Thus the separate lines of such music would then have been more easily and clearly perceived than is possible today when played on most modern instruments with all the problems of balance which they bring with them. See also Appendix 4, 'Alleged Imperfections of Old Woodwind Instruments'.

In addition to the above, the greater definition would encourage the playing of more accurate consonances and heighten their contrast with the dissonances.

The series of pure tones resulting from a single set of simple vibrations whose periods formed an harmonic series (*e.g.* the 'note' of a musical instrument), are sometimes referred to as the '*notes of the harmonic series*' (*of acoustics*), although they are normally perceived blended into a single note. The latter is also true of the so-called '*chord of nature*'—a term fabricated by some musical theorists of the nineteenth century from the fourth, fifth and sixth harmonic upper partial tones (or the first, third and fifth—with suitable octave transpositions) of a single note, and derived from Rameau's observation of the curious parallel existing between these harmonic upper partial tones and the individual notes which form a major triad. Though on first acquaintance the claims of the 'chord' might appear to be genuine, serious investigation shows them to be unfounded and the 'chord' is revealed as a scientific misconception. See also pp. 32, 77 *et seq.*, 110 and Appendix 5.

Both these terms are in reality nothing more than fanciful names for a single musical note.

It is an inevitable arithmetic coincidence that any interval which can be expressed physically as a ratio of two whole numbers must occur between those notes of the harmonic series corresponding to these numbers. But an harmonic musical interval is an effect produced in the hearing faculty by *two separate notes* (*i.e.* two separate sets of harmonic partial tones) *sounded simultaneously*. Its recognition as such is dependent solely on that property of our ears which makes them sensitive to beating and thus has no *direct* connection with arithmetic.

Notes which are not part of the prevailing harmony, such as passing notes, suspensions, or decorating notes, the contrapuntist describes as '*unessential notes*'. Because of their melodic character, such notes are naturally given *melodic intonation*, *i.e.*, an intonation determined largely by the musical instinct of the artist for the essential note to which they are moving.

Notes whose intonation requires adjustment for different concords are said to be *mutable*. Typical examples are the second- and seventh-degree notes of the minor scale (see Fig. D). Such notes are pushed to and fro by consonant intervals, particularly the fifth with its sharp definition. They present no difficulty to unaccompanied voices or strings, etc. The performers have only to sing or play in tune and the notes so sung or played will find their own position. Where and when the ear demands it, the intonation of mutable notes must be *heard* to change while actually being sounded—it is not simply a matter of choice of alternative positions. To instruments of fixed intonation, such as keyboard instruments, however, these notes present a problem which can only be solved by a compromise.

'Notes' are also the signs or marks in staff notation used essentially to exhibit intervals.

15

TUNINGS AND TEMPERAMENTS

Temperament—regular, irregular and cyclic temperaments—equal temperament—effects of inharmonicity on tuning of keyboard strings —(random) inharmonicity—mean-tone temperament—wolf—wolf-fifth—Pythagorean tuning—just temperament—cent—savart.

The sets of intervals produced by tunings are known as temperaments. So that *practical temperaments* and *tunings* are essentially the same things—methods used to tune keyed instruments, particularly keyboard instruments. In the latter case, assuming that a standard keyboard is to be used, then the first practicable step in such a scheme is the elimination of the comma by making both major and minor tones equal.

The notes perceived from any keyboard instrument being limited in number and fixed in pitch make its intonation rigid, and it is this rigidity, rather than the amount of mistuning that must follow, which is the salient feature of any temperament. (But see also para. 3 etc., of Chapter 12.) The extent and distribution of the mistuning is dependent on the musical objective aimed at. The problem of the keyboard instrument is to devise that system of tuning an instrument of fixed intonation which gives the best approximation to the flexible intonation of any given period. But apart from the inherent 'out-of-tuneness' in music performed with such a set of fixed tunings, those with delicate ears and accustomed to the exercise of choice in intonation must detect in these performances a degree of stiffness, a lack of vitality—a certain insipidness in some intervals, an unnatural harshness in others—and find these tunings acceptable only as a makeshift.

What really matters about a temperament is how well it sounds *musically*. The best temperament for the tuning of a keyboard instrument is that which is *judged by the ear* to do the least harm to the music for which it is used to perform. See *regular* temperament, etc.

Temperament must not be confused with scale (see Chapter 16, INTONATION AND SCALES). The intonation of a scale is essentially flexible.

The easiest of all intervals to tune exactly is the unison. The next easiest is the octave, which interval, at least in theory, is tuned exactly in most systems of temperament. Of all the intervals that lie between these two, the easiest to tune exactly—or a little off its exact value—is the fifth (followed closely by its inversion—the fourth). Hence the most practical way of expressing a temperament—whether for the purpose of critical examination or for tuning—is in terms of its fifths.

Any system of tuning in which all the notes can be arranged so as to form a continuous series of fifths of *equal* width is called a *regular* or *linear system* or *temperament*. Conversely, one in which this cannot be done is called an *irregular*, *non-linear* or *unequal* system. Whereas any intervals of the same degree belonging to differing keys of a regular temperament, must, by definition, be of equal width—*i.e.* sound alike—this cannot, by definition, be true of irregular temperaments. For, in the latter case, the intervals between the same degrees of differing keys must vary with the amount and distribution of the particular fifths that it has to employ. (In using regular temperaments it is assumed above that the keys referred to do not include any of those which may lie outside the range for which the temperament was designed to be used.) Thus, in an irregular temperament, subtle differences can exist between keys. Intervals between the same degrees of different keys do not necessarily sound alike. Each key can have a certain uniqueness.

Any set of intervals of the same degree from a system which repeats itself exactly after passing through a certain sequence of these intervals can be displayed in the form of a circle and hence is said to be a *cyclic* or *closed* system. Such a temperament allows endless modulation in these keys. Typical examples are the cycles of 12 (equal temperament), 19, 31 (Huygens), 50 and 53, etc. All these systems are *regular*, *i.e.*, temper the fifth equally throughout, and therefore, from **this** point of view, have just as much claim to the title 'equal temperament'.

Any intervals which occurs in any regular system of temperament may be expressed exactly in terms of the sum or difference of its 'fifths' and true octaves. (In the particular case of the Pythagorean system the fifths are also true.)

E.g. 1 mean-tone major sixth = 3 mean-tone fifths − 1 octave. (See Table A, Chapter 20.)

Any of the intervals which occur in any regular cyclic system of temperament may be expressed exactly in terms of any of the other intervals in the system.

E.g. in the cycle of 31, 23 'minor thirds' = 8 'major sixths'. (See Chapter 20.)

Equal temperament provides the best acoustical approximation, on an instrument of fixed intonation, to the flexible intonation implied in the modern scale-system (see Chapter 16). A good orchestra does **not** play in equal temperament any more than does a bad one.

For even though its keyed instruments are so tuned, their intonation can be adjusted in performance, by at least a comma, at the discretion of the player. The better players do, in fact, **demand** a greater degree of pitch accommodation in their instruments. Thus, so far as the keyed instruments are concerned, equal temperament provides a kind of average intonation from which a certain amount of deviation is possible as and when felt necessary by the musical ear of the player—enabling him to command the widest practicable range of intonation.

Fretted instruments do not necessarily imply the use of a fixed intonation by the player. For instance, on p. 84 of his book *The Instruments of Music* (Ref. FR1) Robert Donington writes, of the viol:—

> The tendency to an edgy tone is reinforced by the pieces of fine gut, called frets, which are always tied round the viol's fingerboard at semitone intervals. Historians have imagined that these are meant to keep the player in tune; and indeed they do serve as a rough guide in that regard. But to play accurately in tune you must use your ears and adjust your fingers (by pulling or pushing) just as meticulously as the violinist: while those who assume that the frets must needs interfere with rapid fingering can have no first-hand knowledge of the correct technique.
>
> The real function of the frets is quite different: it is to give each stopped

> note the same ringing sharpness of quality which only open notes possess on the violin. For on the violin, you stop the string with the soft end of your finger, and get a more rounded tone: on the viol, you press your finger against the fret, which, being hard, gives you a sharper quality of tone....

Again, of the lute he writes, on pp. 90 and 91,

> Like the viol, the lute has frets of fine gut tied round the fingerboard at semitone intervals. These help in catching full chords rapidly and surely: but as on the viol, their real function is to give a sharp and ringing quality to the stopped notes. You cannot wholly trust the frets to keep you properly in tune. For unless you use your fingers nearly as accurately as the violinist does, you may be but indifferently in tune; and your tone, too, will sound lifeless, lacking its proper ring and beauty.

The reason for the success of equal temperament as a tuning for keyboard instruments lies in the historical fact that it made possible the continued use of a keyboard with twelve notes to the octave—such as had previously been used for the mean-tone types of tuning—and, that its particular division of the octave, produced, by a lucky chance, a greater number of tolerable melodic intervals than many other temperaments having more limited harmonic resource. This, however, is not quite the whole story. For, as the author points out in *The Instruments of Music*, p. 243, footnote 1,

> ... it is worth remembering that there is one Renaissance and Baroque keyboard instrument of which the tuning is not fixed rigidly, but is flexible enough to permit just intonation in theory, and with skill perhaps in practice; namely the clavichord. Bach's reputed favourite.

By pressing more or less firmly on the keys the skilled clavichord player can increase or decrease the string's tension about a conveniently chosen average, and is thus able to produce corresponding changes in what could otherwise be a fixed intonation. (See Appendix 4.)

Equal temperament does not, as its name might be taken to imply, temper *all* intervals equally, though a system of temperament that does so—*i.e.* one in which the consonant intervals are made as equally harmonious as possible—was devised some two hundred years ago for the harpsichord by Dr. Robert Smith (1689–1768) working purely from his estimation of the demands of the musical ear. Ll. S. Lloyd brought this tuning to the notice of Dr. A. R. McClure,[1] who

[1] 'Studies in Keyboard Temperament', *Galpin Soc. Journal* I, p. 28. 'An Extended Meantone Organ', *The Organ*, No. 119, Jan. 1951.

after experimenting with it and finding it particularly suitable for the performance of pre-Bach music, started to build an organ to demonstrate it but died before he could finish this work. His organ was, however, completed by Dom Lawrence Bevenot in 1950, and may be seen and played, on application to the Reid School of Music at Edinburgh, where it is now kept. In this instrument 19 notes are made available from what is virtually a division of the octave into 50 equal parts (see Fig. H). A standard keyboard is used giving the following notes, E♭, B♭, F, C, G, D, A, E, B, F♯, C♯, and G♯. By drawing one or more of the 7 key switches provided the notes E♭, B♭, F♯, C♯, G♯, and B and F, may be replaced by D♯, A♯, G♭, A♭, D♭ and C♭ and E♯. See also Chapter 20.

Because the fifth has greater definition than the major third it has been inferred by some that equal temperament tampers most with those intervals which can best stand it. But this ignores the effects of difference tones, and though it might possibly be true of two-part writing it certainly is not so where three or more parts are concerned. See pp. 18 and 19.

Although in theory it would appear that the octaves of all instruments tuned to equal temperament are mathematically true—*i.e.* have a frequency ratio of $\frac{2}{1}$—this is not necessarily so. For instance, the higher the pitch of the upper partial tones perceived from a freely vibrating string which is not perfectly flexible—such as a piano string—the sharper these partials become compared with those corresponding to the harmonic series based on the frequency with which their fundamental partial tone is associated (see *inharmonicity*, Chapter 11). Hence the tuner, who tunes to eliminate beats between the fundamental of the upper note and the second partial tone of the lower note, will naturally stretch the octave *physically*, and the musical ear, not surprisingly, seems to prefer them that way. This is simply a case of the ear asserting its right to judge—within the limits of choice allowed to it—when such an instrument is most nearly in tune with itself *musically*, and provides a good example of the danger of leaving out the ear and regarding any particular musical interval as being defined primarily and absolutely by a *mathematical* ratio.

But this is only one of the effects which inharmonicity has on the tuning. Since intervals less consonant than the octave must necessarily suffer even greater inharmonicity it follows that irregularities in the setting of the bearings (temperament octave) are unavoidable—which, in turn, must result in even greater irregularities in the remainder of the tuning. In fact, most tuners, in an attempt to ease the task of providing—as the scale ascends—a reasonably smooth increase in successive beat-rates for *all* sets of identical tuning intervals, stretch the octave a little past its beatless position. Such a tuning cannot help but be a compromise, which may deviate in either direction from (at least theoretically) exact equal temperament—but is again a case of the ear asserting its preference under very difficult circumstances.

The inharmonicity of a string can be reduced both by increasing its length and by decreasing its diameter. Other things being equal, this reduction is directly proportional to the fourth power of the length and to the square of the diameter. Thus, increasing the length by a $\frac{1}{5}$th, *i.e.* from 1 to $\frac{6}{5}$, reduces the inharmonicity by $(\frac{6}{5})^4 = 2{\cdot}07$, just over two times, *i.e.* more than halves its previous value. Again, decreasing the diameter by a $\frac{1}{2}$, *i.e.* from 1 to $\frac{1}{2}$, reduces its inharmonicity by $(\frac{1}{2})^2 = \frac{1}{4}$, *i.e.* quarters its previous value.

Generally, the inharmonicity of the strings of a keyboard instrument increases in value both upwards and downwards from the middle of the keyboard. Upwards, because of the shortening of the strings, and downwards, because of the thickening of the strings.

All other things being equal, the energy of a vibrating string is directly proportional to its mass, so that the thicker the string the greater the amount of energy that can be taken from it—but, also, the greater its inharmonicity. It is to overcome this latter effect that piano strings are used in unison twos and threes, ('bichords' and 'trichords')—except in the extreme bass—and that all the lower strings are 'wound', 'covered' or 'gimped'.

The coefficient of inharmonicity of the strings of a good harpsichord can reach as low a value as $\frac{1}{30}$th of that of the corresponding strings of a good piano. Thus the effects of inharmonicity on the tuning of such a harpsichord are almost negligible.

If there are, by chance, any soundboard resonances whose

frequencies lie sufficiently close to those of one or more of the sounding strings' partial vibrations as to be excited by them, then a smaller, but not necessarily insignificant form of (*random*) *inharmonicity* can also be introduced. Such inharmonicity can be either positive or negative—depending upon the frequencies involved—*i.e.* can have either a sharpening or a flattening effect on the harmonics of the notes being sounded. However, the larger the soundboard, the greater will be the tendency for its resonances to overlap and hence to neutralise one another's irregularities. The above remarks apply to all stringed instruments. (See also Ref. FR2.)

Pipe organs do not suffer from inharmonicity.

Mean-tone temperament is a type of tuning first standardised in 1523 by Aron. It would solve—for the keyboard instruments of his day—the problem of mutable notes, which would occur in the modes just as they do in the major and minor scales and for the same reasons. Aron's form of this temperament flattened all the fifths and therefore sharpened all the fourths, by $\frac{1}{4}$ comma, and made all the major thirds true. The true major third (ratio $\frac{5}{4}$) is the sum of a major and a minor tone. *Basic mean-tone* temperament therefore made all the whole tones the mean of these two intervals, $\frac{1}{2}$ a comma smaller than the major tone, $\frac{1}{2}$ a comma larger than a minor tone, leaving each of the diatonic semitones (EF and BC) too sharp by $\frac{1}{4}$ comma. The black keys were tuned to give the chromatically altered notes used in modal music: C♯, E♭, F♯, G♯ and B♭. F♯, G♯ and C♯ were tuned to give true major thirds above D, E and A respectively, while B♭ and E♭ were tuned to give true major thirds below D and G. The result of tuning the black notes in this way was to produce 'wolves' (see p. 169) in the keys E♭ major and E major, and in all the major keys more remote, while the minor keys of C and E were faulty. This was why on some organs, such as that built by Father Smith for the Temple Church, the black notes for E♭ and G♯ were divided across the middle, and the back half raised—additional ranks of pipes being inserted in each octave to correspond with the notes so added for D♯ and A♭.

In the days when mean-tone was in common use, keyboard players tuned (and retuned) their instruments to suit the music they were about to play.

The 'central' key was not necessarily C. The compass of available keys of an instrument would be moved 'sharpwards' or 'flatwards' according to the demands of the music.

Further, even assuming such a keyboard player to have been taught initially to tune to a certain theoretical temperament, he would almost inevitably, in the process of time, acquire certain preferences in the width of his intervals—born of the desires and experiences of his musical ear—which, together with a practical realisation of what could and could not be gratefully accommodated in an instrument of fixed intonation, would cause his normal tuning to deviate appreciably from its theoretical origin.

Of Bach's tuning we are told, by Philip Spitta,[1] '. . . . That he evolved all this by his own study and reflection, and not from reading theoretical treatises, would be very certain, even if we had not the testimony of his contemporaries; and he carried out his method with such rapidity and certainty that it never took him more than a quarter of an hour to tune a harpsichord or clavichord.' Thus it would appear that Bach knew exactly what kind of temperament was best suited to his music, and also how to tune it to his satisfaction with speed and efficiency. It is generally acknowledged today that the tuning in which Bach was content to play all his harpsichord and clavichord compositions was not that which we now call equal temperament, but was an irregular form of cyclic temperament requiring the thirds to be sharpened in various degrees.[2]

So long as composers restricted themselves to a few closely related keys and made little use of chromatic harmony, the sweet thirds of the mean-tone types of tuning prevailed over the theoretical tidiness of equal temperament. But gradually, as these restrictions were relaxed, the tuning of the major thirds was forced more and more sharpwards until it met the requirements of equal temperament. Thus were the 'wolves' (see p. 100 and Appendix 7) of mean-tone divided and dispersed equally throughout those keys which have now come to be associated with our music in general, and with our standard

[1] *Johann Sebastian Bach*, Vol. 2, p. 42, Eng. Trans., 1884-85.

[2] See, for instance, a paper by John Barnes entitled *Bach's Keyboard Temperament: Internal Evidence from the "Well Tempered Clavier"*, Early Music, Vol. 7, No. 1, Jan. 1979.

keyboard in particular. But there was a long and sometimes bitter struggle over the years before the matter was finally (?) settled, and during this intervening period many different forms of intermediate tunings of both regular and irregular types were brought into use.

For details of the technique employed in tuning a large number of these temperaments see Owen Jorgensen's book *Tuning the Historical Temperaments by Ear* (Ref. FR3).

If, in basic mean-tone tuning G♯ (772·6) was substituted for A♭ (813·7 cents) the result was a badly tuned interval (about 2 commas out) or '*wolf*'. The normal tuning was from E♭ (310·3 cents) through 11 mean-tone fifths to G♯ (772·6). This made the interval G♯ E♭ nearly 2 commas (about 40 cents) sharp of a mean-tone fifth—hence its name the *wolf-fifth*. (See also Appendix 7.)

Just temperament differs in detail, but not essentially in nature, from other temperaments. The seven notes of its cycle give perfectly tuned intervals for major triads on the tonic, dominant and subdominant, and their inversions, and although this tuning does make most of the diatonic intervals just (*i.e.* true), it undoubtedly tempers two of them by as much as a comma—those between the second and fourth, and second and sixth notes—and is therefore not a scale, but a temperament (though not a practical one). Moreover, the tonic relationship which this cycle implies is not an essential element of the structure of our musical scale. Witness, for example, the evidence of the earlier (modal) music of Western Europe. Again, suitability for the production of melodic intervals—such as would be required by 'unessential notes' *etc.*,—is not even considered. So limited, in fact, is the musical objective of this cycle that it is suitable only for such things as the two forms of Amen which occur in the Liturgy, and some of their decorated equivalents. Even the Tudor musicians were well aware that this tuning could not be used in real music.

However, because the positions of its notes are fixed by consonant intervals (*i.e.* can be stated *exactly* in terms of ratios of low numbers) and are therefore, at least in theory, tunable *exactly*, and because most of the intervals formed have a more or less simple relationship with the tonic, this temperament is often used to establish standard reference points from which to estimate and compare, physically—*i.e.* in terms of vibrations alone—the degree of mistuning of the intervals produced by different kinds of temperaments. See Table A, Chapter 20, where this tuning has been extended in both directions—as one might do with a regular temperament—and pages 63 and 64, where it has been used to estimate the interval by which twelve true fifths exceed seven true octaves.

A *scale* is a community of *intervals*, not of notes.

'Just intonation' must not be confused with 'just temperament' (see Chapter 16, INTONATION AND SCALES).

Pythagorean tuning is a system of tuning in true fifths as used by the strings of the orchestra. Since string players prefer to avoid the use of open strings, such tuning does not in any way limit their choice of intonation.

If this tuning is extended to include seven (or more) successive fifths, tuned upwards, and the notes so 'generated' reduced to the same octave, the result is a Pythagorean 'scale'. It would appear that the scale used in plainsong and by strong players in unaccompanied melody tends to approximate to this type of intonation.

The *cent* is a useful logarithmic unit of ratio in which to express the physical errors of tuning. As is the case with all logarithmic units, it gives a direct measure of ratio. The musical interval to which one cent corresponds is very small—one hundredth part of an equally tempered semitone. As such it has no practical significance in musical performance. Beats between fundamentals of a note pitched at middle C and another pitched one cent sharper would occur less than once in $6\frac{1}{4}$ seconds. Its equivalent frequency ratio is roughly 1·0006 to 1 [1·000576... / 1], or a frequency change of about ·06 per cent.

1 semitone of equal temperament = 100 cents.

1 octave = 1200 cents.

The number of cents corresponding to any given ratio may be obtained, approximately, by multiplying the common logarithm of the ratio by 4000—or if much greater accuracy is required—by 3986·3137 (see SUMMARY OF MATHEMATICAL TERMS, p. 254).

The *savart* is a similar unit to the cent but is about four times as large. Its equivalent frequency ratio is roughly 1·0023 to 1 or $\frac{436}{435}$.

1 semitone of equal temperament = 25·08583 savarts.

1 octave = 301·0300 savarts (see SUMMARY OF MATHEMATICAL TERMS, p. 254).

1 savart = 4 cents (more accurately 3·986314 cents).

Conversely, 1 cent equals about a quarter [·2508583] of a savart.

The number of savarts corresponding to any given ratio may be readily obtained by multiplying the common logarithm of the ratio by 1000.

16

INTONATION AND SCALES

Scale and scale-system—pure scale—intonation—'in tune'—just (or true) intonation—diatonic scale or mode—enharmonic change—hexachord.

Composers do not think, nor do skilled artists play, in a scale if by a scale is meant a series of fixed intervals—more particularly if these intervals are thought to originate solely as relations between sets of *adjacent* sounds, arranged in order of pitch, such as might be suggested by a run of notes from a keyboard instrument. The composer simply writes music. To discover the nature of the material he uses his scale must be distilled from the music he writes. Failure to realise this has been responsible for more mistaken theory about music than anything else. *Scales are derived from music as performed*—and *not vice-versa.* A *scale*, or more appropriately, a *scale-system*, is merely a method of classifying and labelling the musical material used by composers and skilled artists. It must not be thought of as a series of notes bearing a fixed relationship to a tonic, but as a fluid and well ordered community of intervals, a community which acquires, and in turn loses certain patterns as classical tonality enters or leaves it at the will of the composers. Thus, the *pure-scale* or *pure scale-system* may be defined as one in which the essential intervals are in tune and the intonation is flexible, the less significant of these intervals being liable to be compressed or expanded under the harmonic stress of the more significant ones. The latter applies especially to the fifth, though its exceptionally strong influence as a melodic interval should not be overlooked.

This pure scale-system was distilled by Stanford from his study of modal counterpoint of the sixteenth century. Its basic principle is that diatonic semitones are fixed intervals, while major and minor tones are changeable. This principle, however, applies only to those intervals produced in concords. So that, for instance, *it is only when the leading note is sounded with the dominant, as a major third, in a concord, that it should form a diatonic semitone below the tonic.*

Such a scale can never be produced by a keyboard instrument unless the aural perception of the trained musical ear enables it to hear the fixed tuning as something which can deviate from its physical origin. The available evidence suggests that many people hear the piano in this way.

Good *intonation* is concerned with that *choice* of intervals—melodic or harmonic—*judged solely by the ear* to best suit the musical occasion *as it arises during performance.* No study of intonation is complete which does not take *all* these factors into account.

The flexibility of the scale is demanded first, in order that consonant intervals may be *maintained at their correct size* whatever their position in the scale, and secondly, so that melodic intervals, and those included in enharmonic change (see below), may be allowed to *vary in size* in accordance with the demands of the musical circumstances. These demands are, in turn, based on the following fundamental facts:—

(1) Successive tunings in fifths, or in major thirds, can never exactly coincide with successive tunings in octaves which begin with the same note. Thus some notes of the scale must be mutable—even when no modulation is required.

(2) Modulation, for the same reason as above, requires further notes to be mutable.

(3) The intonation of decorating notes is decided largely by the influence of the melodic line on the feelings of the accomplished artist. As long as he produces the *melodic* intervals of *his* choice, he cannot, strictly speaking, be accused of not playing or singing '*in tune*'. Whether, in the circumstances, his choice is the same as ours would be, is another matter. Only in the intonation of consonant (harmonic) intervals—which includes unisons—does singing or playing 'out of tune' have any definite meaning. It is, for instance, presumptuous nonsense to

suppose that Eastern music is 'out of tune' simply because it uses some melodic intervals which may be unfamiliar to Western ears. See also the remarks on 'passing notes', p. 45.

No good accompaniment doubles the melody.

Except when it lasts for only a moment, a skilled string or wind player, or unaccompanied choral singer, gives perfection of consonance—*i.e.* playing or singing 'in tune'—an importance which entirely overwhelms any desire for fixed intonation.

The origin of our scale can be traced back to modal music. The essential element of the scale of modal music was the hexachord (see p. 179). It was far more than merely a method of sight-reading. It was the basis on which many of the plainsong melodies were constructed—whatever their mode. Further, this continued to be so as long as melodic lines remained the most important consideration, as in the concurrent melodies of polyphony that followed. It was fortunate for European music that the scale of plainsong—which almost certainly employed Pythagorean intonation, or at least something approaching it—lent itself so readily to development in this direction. The flexible scale of unaccompanied polyphonic music came into being in this way. The compromise of fixed tuning (temperament), was imposed later by organs and other keyboard instruments in an attempt to imitate—as best they could—the flexible intonation demanded by the ears of the unaccompanied singers.

The oft quoted statement that 'music and mathematics have an association which goes back at least to the time of Pythagoras' is not true, though it does become so if the word 'music' is replaced by 'the *tuning* of musical instruments'. For the problem of tuning to a temperament is essentially one in physical acoustics and has an arithmetical basis because correctly tempered tunings are concerned with the estimation of *physical beats*—which must, in turn, depend on the estimation of *rates of vibration*. It follows that an arithmetic approach is appropriate in the discussion of tempered tunings intended to represent, as best they may on a keyboard instrument, the flexible intonation employed in any particular period. But when investigating, in its scientific aspect, the scale implied in *music* as conceived by the composer, and rendered by a good artist on an instrument having free intonation, then the problem is one in the physiological and psychological sections of *hearing*—as was insisted by Helmholtz—and not one of arithmetic relationships.

So far as musical ability is concerned, most of the work done so far seems to indicate that mathematicians are less gifted than are non-mathematicians.

The 'just intonation' of the theoretician is a theoretical conception of the scale with fixed intonation, that gives, with mathematical truth, the ratios required to produce tonic, dominant and subdominant concords (see TUNINGS AND TEMPERAMENTS, *just temperament*). To the musician, on the other hand, '*just* or *true intonation*' means the playing or singing of the intervals of *concords* in tune with ideal accuracy—as and when they occur in the *performance* of music; and this requires a scale with flexible intonation for concords which produce mutable notes. It does *not* imply an intonation which makes *all* intervals perfectly true *regardless of the musical circumstances*. See also flexibility of scale on previous pages, and *comma* (*of Didymus*) in Chapter 12.

Any set of intervals formed by sounding successively the 'white notes' of a keyboard instrument may be regarded as constituting—as near as the rigid tuning and temperament of such an instrument will allow—the musical material used in a *diatonic scale* or *mode* of one form or another. Thus all the Church modes as well as the Ionian (scale of C major) must be considered as falling into this category. The same is obviously true of any set combination of 'black and white notes' which can be made to conform to the above after suitable transformation. However, tonality (or modality) is not established simply by the choice of a particular set of intervals. It is also concerned with the manner in which these intervals are used. For instance, though both the Dorian and Hypomixolydian modes make use of the range of intervals from D to D', in the first of these modes intervals which make use of D and A are favoured, while in the second preference is given to those intervals which include G and C.

One way of effecting a modulation is to approach a note or chord in one sense and to leave it in another. This necessitates a small shift in intonation (actual or implied) known as an *enharmonic change*. Because of their unstable intonation, such

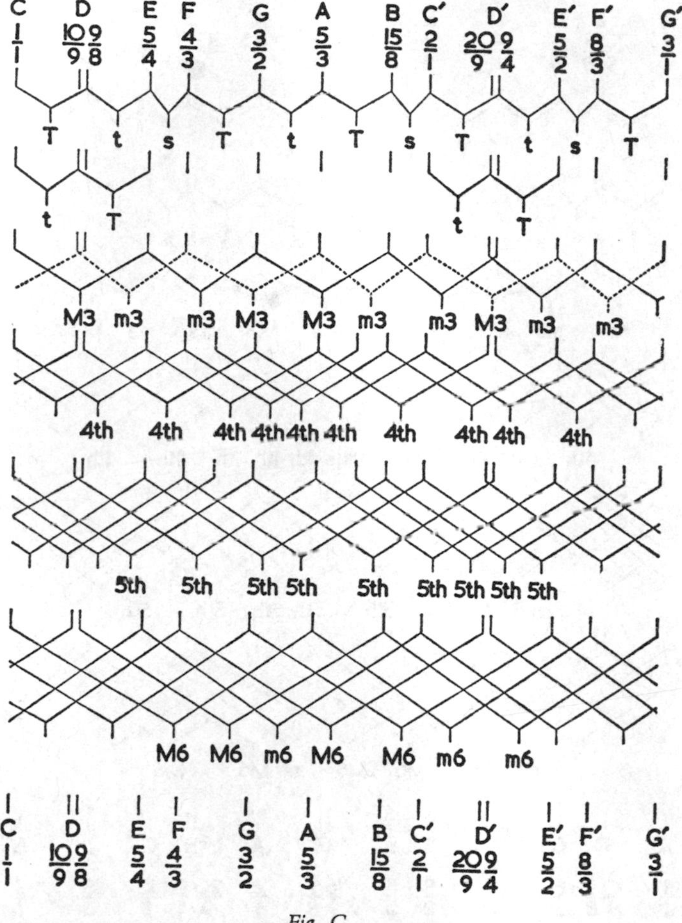

Fig. C

changes are best made in discords such as the diminished seventh or minor ninth, where, for instance, in passing from one key relationship to another, a diminished fifth, bounded by—say—a C and G♭, may be changed—or at least be perceived to change mysteriously—into an augmented fourth bounded now by a C and an F♯. Enharmonic changes can be, and are in fact actually made by voices and all instruments in which sufficient adjustment in intonation is possible, and although the piano, with its limited five black notes, cannot

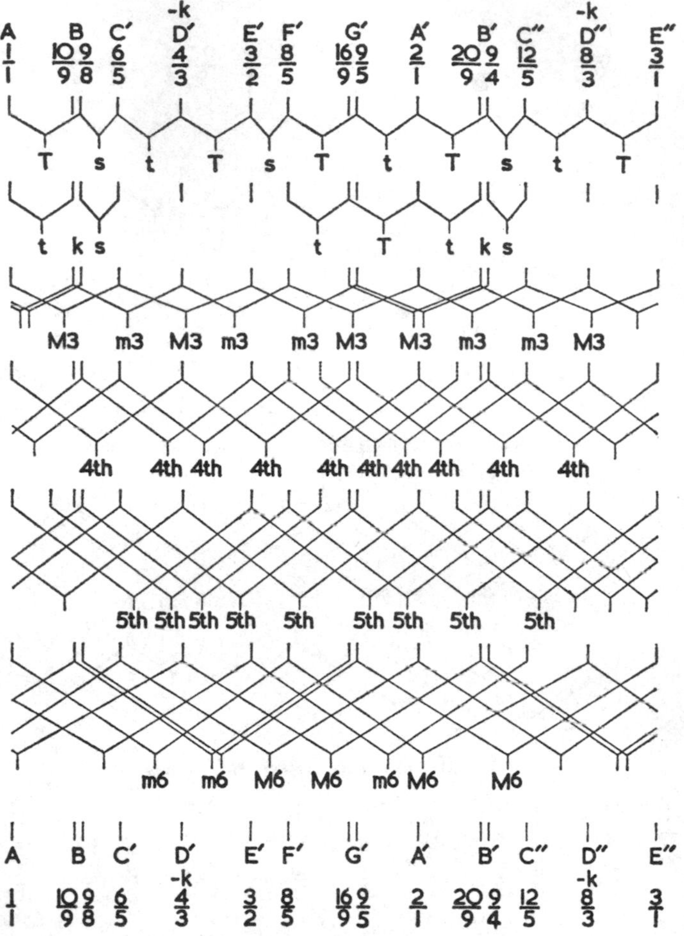

Fig. D

provide the necessary physical change, nevertheless, the ear can still perceive from its notes some degree of enharmonic change whenever the unexpected resolution appears.

As soon as a player (or singer) of concerted music becomes aware of the dissonance of a mistuned consonant interval his sensitive ear compels him to make all the small changes in the width of his intervals which it judges to be necessary—to the extent demanded by his aural perception and permitted by the range of flexiblity of intonation possible on his instrument.

For instance, an enharmonic change which is very perceptible to the sensitive ear of such an artist would call for a change of intonation by some 20 cents.

Enharmonic change has been supposed by some people to have originated with equal temperament. It does, in fact, presuppose true intonation and dates back to polyphonic times.

The name *hexachord* is given to any group of six consecutive intervals of the diatonic scale—in particular those commencing on the tonic, dominant and subdominant. These particular (major) hexachords are identical in all respects so far as their intervals are concerned (provided that B♭ is allowed to replace B in the hexachord on F), and each one equals the greatest span of consecutive intervals reproducible exactly at two other pitches in the same diatonic scale. Further, these hexachords contain all the consonant intervals less than an octave.

The tritone is possible only in a hexachord on D where it can be avoided by calling again on B♭ to replace B.

Figs. C and D are intended to show graphically some of the most obvious limitations of just intonation, with its set of fixed tunings tied to a tonic. Things such as melodic intonation, modulation, use of enharmonic change, etc., which must make practically every note mutable at some time or another, and which must depend upon the particular musical circumstances, cannot possibly be taken into account in a diagram. It follows that any practical usefulness of these diagrams is limited to giving some idea of the state of affairs existing in the major (Fig. C) and minor (Fig. D) scales—*so far as consonant (i.e. harmonic) intervals only are concerned—during performances in which the barest minimum of mutable notes are demanded by the music.*

Many primitive flutes and pipes have been found to have equidistant finger-holes along their lengths, and some writers, in an attempt to justify their theories of primitive melodic scales, have assumed that equal intervals would result from equal changes in the effective lengths of such flutes and pipes as would be brought about by stopping or unstopping adjacent holes in them. In fact, in such a case no two intervals would be the same; and this is also true of the intervals which would result from equal changes in the sounding-lengths of vibrating strings. The frequencies of vibration of both air-columns and strings vary inversely as their lengths.

17

THE CONSTRUCTION AND CALIBRATION OF A SIMPLE MONOCHORD

Although good intonation is a matter of intent listening and control—a largely personal affair and one which is quite independent of numbers—nevertheless, if a thorough knowledge and understanding of intervals is to be acquired, it is essential to be able to record one's choice in this matter and to relate it to that of others by the use of some convenient standard, some form of common currency. Nowadays intervals are usually quoted in terms of the frequency ratios to which they correspond—and although, in general, frequency is a difficult quantity to measure accurately without elaborate equipment, fortunately, the frequency of a vibrating string is inversely proportional to its length so that the interval corresponding to any (frequency) ratio can be formed by simply sounding string-lengths whose ratio equals the inversion of the required (frequency) ratio. Thus in order to sound the interval of, say, a major third (which corresponds to a frequency ratio of $\frac{5}{4}$ or 5/4), string-lengths of ratio 4 to 5 or 4/5 would need to be sounded.

The monochord is not, in the generally accepted sense of the word, a musical instrument. It cannot reproduce the warm, live, dynamic intonation obtained in musical performance by a good artist using an instrument which allows him to exercise some choice in the width of his intervals. In this sense its intonation is static. Nevertheless the monochord provides the simplest foundation on which a critical study of musical intervals can be built.

In the first edition of his '*Tonempfindungen*' Helmholtz writes on page 499:—

'Down to the seventeenth century singers were practised by the monochord, for which Zarlino in the middle of the sixteenth century reintroduced the correct natural intonation. Singers were then practised with a degree of care which we have at present no conception. We can still now see from the Italian church music of the 15th and 16th centuries that it was reduced to the purest euphony of the consonances and that its whole effect is destroyed as soon as these are executed with insufficient purity.'

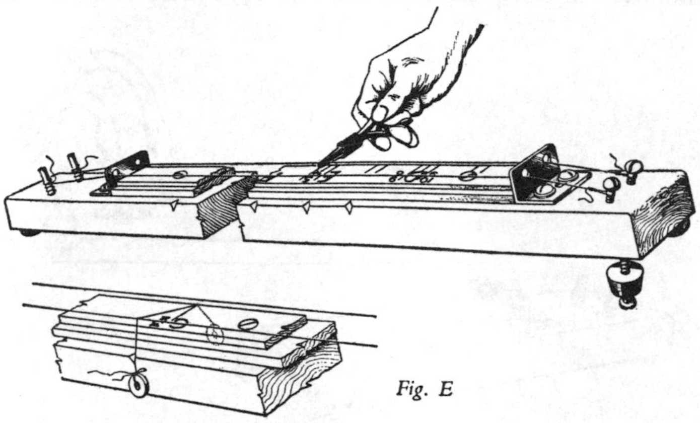

Fig. E

While in the chapter headed 'Retrospect' he writes:—

'A man who has never made physical experiments has never in the whole course of his life had the slightest opportunity of knowing anything about pitch numbers (frequencies of musical vibrations) or their ratios. And almost everyone who delights in music remains in this state of ignorance from birth to death.'

The monochord illustrated in Fig. E has been designed with the main emphasis on simplicity of construction—the object being to encourage as many people as possible to make one and carry out the experiments described in Chapter 18.

The base of the monochord consists of a piece of 2″ × 1″ seasoned wood of length 48″ to 53″—the greater length being required if a diaphragm-type stethoscope or a contact microphone is required to be used with it. The two bridges consist of metal angle pieces, fixed very firmly to the wooden base with screws, and having two holes through them to take the wires—or 'strings'. The inside faces of these bridges are spaced apart by a piece of wood one metre long—such as a standard metre scale. The 'strings' consist of steel music wire of about

·008″ diameter. The small diameter virtually eliminates any inharmonicity in the lower partial vibrations of the wire when vibrating freely, and thus enables a reasonably well defined note to be produced by plucking—but this is only achieved at the cost of shorter duration and lower intensity as compared with thicker wires. To partly compensate for the latter, particularly in noisy surroundings, it is recommended that a

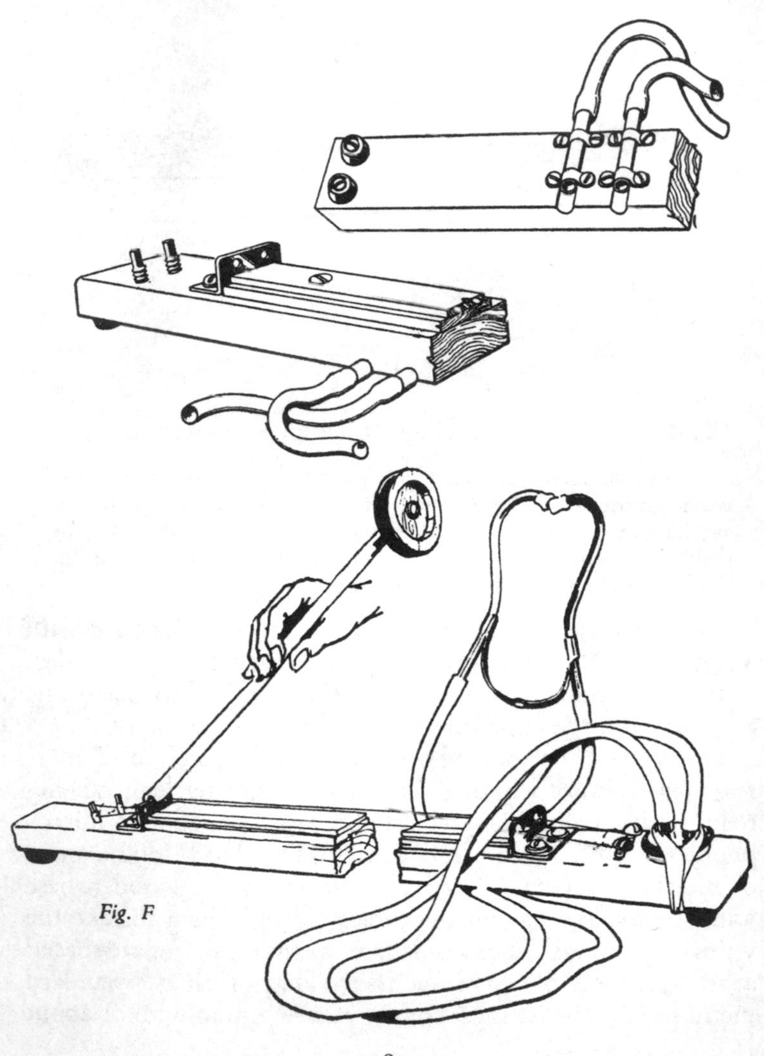

Fig. F

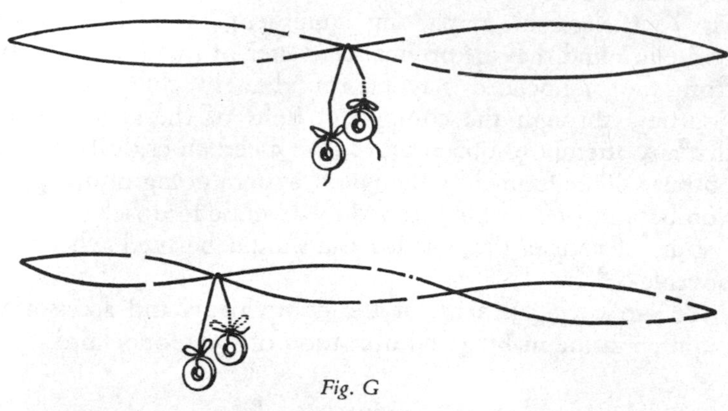

Fig. G

stethoscope be used in conjunction with the monochord. Fig. F shows two different types of stethoscope being used with the monochord. One is simply a length of wooden rod ('dowelling') screwed to a circular-shaped wooden block of the kind that is sometimes used to mount electrical fittings on—the other is a standard diaphragm-type stethoscope, as used by G.P.s, with its diaphragm held firmly against the top surface of the wooden base by means of an elastic band. The first type has the advantage of simplicity and cheapness, the second, freedom of movement while listening, and a better note-noise ratio—as it not only effectively increases the intensity of the sound from the monochord but at the same time considerably reduces the intensity of any 'stray' sounds that may be coming from elsewhere. Two short pieces of metal tubing, held by metal saddle clips screwed to the wooden base, may be used instead of a diaphragm head. This method gives results equalling those using the diaphragm type head, costs much less, is more permanent, and can easily be repeated at convenient distances along the base board thus making it possible for more than one person to 'listen in' at the same time—an obvious advantage when using the monochord for teaching. Alternatively, a microphone, preferably of the contact type, can be used in conjunction with an amplifier and loud-speaker—such as is incorporated in a tape recorder. When using this system the microphone should be placed in the same position as that of the stethoscope diaphragm in Fig. F. If electronic amplifying equipment is used it should be borne in mind that the original character of the sound energy from the monochord may be considerably 'coloured' in its 'journey' through the component links of the system—and that any attempt to obtain more than a certain critical level of output will be limited by the whole system going into oscillation because of instability caused by acoustic feedback.

But, of course, the unaided ear should be used whenever possible.

The following is a list of the components and accessories required in the making and operation of the monochord.

Components

1. Wooden base 2″ × 1″ × 48″ to 53″.
2. Wooden packing piece.
3. Metre-length wooden scale or strip.
4. Three countersunk-head wood screws for fixing items 3 and 2 to item 1.
5. Two steel angle pieces for bridges.
6. Four round-head wood screws and washers for fixing items 5 to item 1.
7. Two harpsichord, or zither wrest pins.
8. Two hitch pins for anchoring strings.
9. Two lengths of steel music wire, 00 Music Wire Gauge, 35 S.W.G., ·008″ or ·213 mm diameter.
10. Four rubber feet.
11. Four round-head screws for fixing items 10 to item 1.

Accessories

12. A pair of round-nosed jeweller's pliers.
13. A tuning key to fit item 7.
14. A stethoscope.
15. Node locators.
16. A plectrum.

Notes Regarding These Items

Item 3. If a standard metre scale is used be sure that it is one metre between the *extreme* ends. Some scales have a little extra at each end for protection. Calibrated metre scales may be obtained from suppliers of educational or scientific equipment.

If a scale calibrated directly in frequency ratios is preferred then a similar piece of wood is required but without metric calibrations. This piece of wood should first be placed on a flat surface and clamped alongside a standard calibrated metre scale. The lengths corresponding to any frequency ratio can then be transferred from the calibrated to the uncalibrated scale by means of a sharp penknife or scriber and a toolmaker's or carpenter's square. A list of ratios and their corresponding string-lengths is given at the end of this book, though the simpler ratios can be easily worked out. Thus, since the frequency ratio for a major third is $\frac{5}{4}$, it follows that the required length will be $^4/_5$ of 1000 mm, *i.e.* 800 mm; that for a major sixth (ratio $\frac{5}{3}$), $^3/_5$ of 1000 mm, *i.e.* 600 mm; an octave (ratio $\frac{2}{1}$), $^1/_2$ of 1000 mm, *i.e.* 500 mm, etc.

Item 5. The bridges illustrated in Fig. E are in fact mass produced by Thomas Crompton of Makerfield, Lancashire. They are known commercially as 'Table Stretcher plates (angle type) No. 315'—and are generally obtainable from ironmongers, etc. Care should be taken to mount these brackets squarely—*i.e.* at right-angles to the metre-scale.

If facilities are available for making these bridges individually, then, by making the holes a little lower the packing piece (item 2) can be omitted. The correct height for the holes is that which allows just sufficient gap between the string and the scale to clear the diameter of one of the jaws of the round-nosed pliers (item 12). By so doing no appreciable deflection or additional tension is introduced when using the pliers to stop the string.

It may be necessary to file that part of the hole in the bridges in which the string lies in order to ensure that the edge of the hole nearest the calibrated scale does in fact stop the string—since if the back edge is slightly higher than the front then the actual length of the sounding portion of the string will be more than the distance between the inside faces of the bridges (*i.e.* the length of the calibrated scale) by one, or possibly two thicknesses of the metal forming the bridge.

Item 2. This should be a little shorter than the calibrated scale—so as to leave the bridges spaced apart solely by the scale.

Its thickness (or height) should be such as to give the correct height between the bottom of the holes of the metal bridges and the string (see item 5 immediately above). The width does not much matter though it should be at least equal to that of the calibrated scale.

Item 6. In the model shown in Fig. E, four brass, round-head, No. 8 gauge, wood screws, one-inch in length were used, together with four brass 2 B.A. washers.

Item 7. The holes for these wrest pins should be made at a slight angle to the vertical so that with the pin in position there is no tendency for the wire to slide up the pin during adjustment. Each pin must be a tight fit in its hole. It should be hammered home.

Item 8. Two screws as used for item 6 will do.

Item 12. A suitable type is made by Brindleys and may be purchased from Buck & Ryan Ltd., 101 Tottenham Court Road, London, W.1.

Item 13. Clock keys are made which will fit item 7 but finer adjustment is possible using a standard harpsichord or zither tuning key.

Item 14. The simple type has already been described. The type used by G.P.s can be obtained from any supplier of surgical instruments etc.

In the tube-type stethoscope head, illustrated in Fig. F, the tubes are of copper, about 2″ long, 1/4″ dia. bore and 3/8″ outside diameter, the saddle clips 'Pyrotenax Ref. 310-369' and the fixing screws brass, round head, 1/2″ long, No. 8 gauge. In this particular case, in order to make a satisfactory job, it was found necessary to let the two tubes a little into the wood, as shown in the sketch, by gouging slight grooves in the base board. This latter operation also helped to prevent the tubes touching the surface on which the rubber feet were standing.

Item 15. This item has a sketch to itself in Fig. E where it is shown placed across one of the strings of the monochord. The node locator consists of a suitable length of cotton the ends of which are tied to two small weights—in this case two 2 B.A. nuts were used as they happened to be handy. So far as the author is aware this is its first appearance. To prevent the node locator from moving out of its set position when in use fine saw-cuts should be made, as shown in the sketch, on both sides of the wooden base at lengths of exactly $^1/_2$, $^1/_3$, $^1/_4$, $^1/_5$ and $^1/_6$ of the 1000 mm scale. These slots must not be so narrow as to jam the cotton. See also Fig. G.

Items 7, 9 and 12 may all be purchased from Fletcher & Newman Ltd., 39 Shelton Street, London, W.C.2.

For those who like to experiment many modifications are possible. To mention only two:—a third string may be fitted between the other two—guitar machine heads may be fitted instead of wrest pins, etc.

Checking the Calibration

Having calibrated the monochord *geometrically* it is a good idea to make some *aural* checks on it. Such checks are possible at certain salient points, namely, those points at which the divided portions of the string, when sounded, give consonant intervals. For instance, at the point marked 5/4 the string will in fact be divided into lengths of $^4/_5$ and $^1/_5$, a ratio of $\frac{4}{1}$, which, when sounded, will give a double octave; at the point marked 3/2 the two string-lengths will be $^2/_3$ and $^1/_3$—a ratio of $\frac{2}{1}$ which will sound an octave; at the point marked 2/1 the two string-lengths will be $^1/_2$ and $^1/_2$—a ratio of $\frac{1}{1}$ giving a unison; and so on.

The required division can be made by either pinching the string with the round-nosed pliers and then plucking it on either side of the point, or alternatively, by tapping fairly softly at the point with, say, the nose of the pliers (better still a blunt knife edge), taking care not to lose contact with the string in the process.

The first method gives much louder results than the second, but by repeating the second rapidly enough an almost continuous adjustment of pitch is possible enabling the required point to be 'felt out' by passing backwards and forwards through it a few times. The latter method is of course virtually a 'clavichord' action. The centre of the string can be found with surprising accuracy by this means.

Since the above notes were first written an entirely plastic stethoscope has become available (*e.g.* the Stethoclip—as supplied by Amplivox of England). This is much lighter and more comfortable to wear than is the metal type, and can be used by simply inserting its end into a hole in the wooden base of the monochord. No pick-up head is then required, but the hole should be a reasonably good fit on the tube and should be open at the far end, *i.e.* the hole should be drilled right through the wooden base.

18

VARIOUS EXPERIMENTAL AURAL TESTS ON THE MONOCHORD

In the directions which follow, it is assumed that the effects of inharmonicity on the string's lower partials can be ignored. This is more or less true of the monochord described in the previous chapter.

Also, that all stopping of the strings is assumed to be carried out using the round-nosed pliers, and for the sake of brevity, 'at $^1/_3$, $^1/_5$, etc.' is used to mean 'at one-third, one-fifth, etc. of the total length (1000 mm) of the string'—measured from either end.

The following tests are best carried out in very quiet surroundings—such as might be found in a library after closing time.

I

Object. To produce in the ear, one at a time, the upper partial tones or overtones of a vibrating string, by eliminating, as far as possible, all partial vibrations other than that corresponding to the one particular overtone chosen.

Hence to use this technique to tune a string to any given frequency, (e) and to tune two strings to an accurate unison (f).

(a) Tune the 1st string to any convenient pitch. Place a node-locator at 1/2 and pluck at 1/4. This will produce in the ear the maximum amount of 2nd partial tone (or 1st overtone) from the string.

Strictly speaking, there will also be produced some 6th, 10th, 14th, etc. upper partial tones (*i.e.* all the odd partial tones of half of the string's length), but these will be weak compared with the 2nd.

(b) Move the node-locator to 1/3 and repeat as above plucking at 1/6. This will produce the maximum amount of 3rd partial tone (or 2nd overtone).

(c) Now place the locator at 1/4, 1/5, 1/6, in turn, plucking at 1/8, 1/10, and 1/12 respectively. This will produce, in turn, in the ear, the 4th, 5th and 6th partial tones.

(d) Thus to generate the maximum n-th partial vibration the node-locator must be placed at 1/n, and the string plucked at 1/2n.

An illustration of the manner in which the string vibrate when a node-locator is placed at $\frac{1}{2}$ and $\frac{1}{3}$ is shown in Fig. G—the amount of deflection being here greatly exaggerated to show the point.

Although much higher partials than those mentioned above can be and certainly are generated and heard—particularly as tone-colour—for the purposes of these simple tests they are not of great importance. With the monochord described in this book these higher overtones progressively weaken, shorten in duration, and become more difficult to isolate satisfactorily by this simple but relatively insensitive method.

The open monochord string can be tuned to any frequency up to about a full tone above 'a' (220 cycles per second). The material and dimensions of the string, as recommended, are such that the tension required to raise the frequency higher than this is sufficient to break the string. On the other hand, when the tension is too low the note sounded is musically of a poor quality.

(e) To tune the open string using a tuning fork.

(1) Strike the fork and place its stem on the top of the wooden base of the monochord. As soon as possible after this, pluck the string at 1/2 and adjust its tension until no difference in pitch can be detected between the two notes sounded.

This test is easier to carry out if the stem of the fork is firmly fixed to the base of the monochord by a strong clamp, or better still, screwed and secured to it by a nut and washer.

(2) Continue as at (1) but now switch the attention of the ear from pitch difference and concentrate on and eliminate any beats occurring between the notes produced by the

combination of the fork and the fundamental or 1st partial vibration of the string. These beats can often be *felt* by lightly touching the wooden base of the monochord with the finger.

(3) If no fork of a frequency low enough to tune the fundamental or 1st partial of the string is available, then the string may be tuned by placing a node-locator at 1/2, plucking the string at 1/4, and then tuning, as at (1) and (2) above, but to the 2nd partial vibration of the string. In this manner it is quite easy, using a standard 'a''—440 cycles per second—fork to tune the string so as to sound 'a'—220 cyles per second—in its open condition, *i.e.* with the node-locator removed.

Since beats always occur at the difference frequency, it is possible, if so desired, to tune a little either side of the fork's frequency—the maximum possible amount of such deviation being limited by the speed and accuracy with which the ear can detect and count the beats. If, for instance, in the above case, 4 beats per second were heard, then this would mean that the frequency of the 2nd partial vibration of the string was either 440 + 4, *i.e.* 444, or 440 − 4, *i.e.* 436 cycles per second, thus making the fundamental frequency either 222 or 218 cycles per second.

If, in such a case, the ear cannot tell from its sense of pitch which is the higher or lower of the two notes, then the best method of deciding this is to adjust the tension of the string so as to approach, attain, or pass right through the condition of unison. Increasing the tension after a condition of unison has been reached must increase the frequency and hence sharpen the note, whereas decreasing the tension—from the unison condition—must result in a decrease of frequency and hence a flattening of the note produced.

The duration of the note is very important here. The greater this is the larger is the number of beats that it is possible to count in any particular case, hence the more accurate is the figure obtainable for the number of beats per second—as calculated by dividing the total number of beats by the number of seconds during which they occur.

Another excellent way of singling out the partial vibrations of a complex periodic vibration, one at a time, is by holding the vibrating body of an electric razor—preferably of the vibrating type—against the face in the hollow of the cheek and adjusting the size of the opening of the mouth and the tension in the muscles of the jaw. By careful adjustment harmonic partial vibrations from about the 3rd to about the 12th can be sorted out and their corresponding notes clearly heard.

Any similar vibrating body would do equally well.

(f) In order to sound harmonic intervals, and to carry out certain other tests, a second string, tuned in unison with the first, is required. For those who are not used to tuning unisons, the following procedure—using the technique described above—is recommended.

(1) Pluck each string alternately at 1/2, adjusting the tension in the 2nd until no difference in pitch can be detected.

(2) Place a node-locator across both strings at 1/2, pluck at 1/4, and tune for no difference in pitch. If, or when no difference in pitch can be detected, sound the two strings simultaneously and tune for absence of beating between the 2nd partial tones.

(3) Next place the node-locator, in turn, across both strings at 1/3, 1/4, 1/5, plucking at 1/6, 1/8, 1/10 respectively—in each case repeating the appropriate procedure as at (2) above.

Alternatively, tuning may be done *without* the node-locator, which can then be used *after* tuning as a test of skill in tuning without its aid.

Thus, assuming that partials of sufficient number, loudness and duration are present, then these beats provide, as the series is ascended, a progressively finer vernier adjustment of the unison condition.

II

OBJECT. To show that, in general, plucking a string at any given point gives a maximum to those partial vibrations for which the point is an antinode (point of maximum movement) and eliminates all those partial vibrations for which the point would be a node (point of no movement).

Thus, to take a particular case, although a string cannot, using the technique of section I, be made to sound only its fundamental (or 1st partial tone), nevertheless, if plucked exactly at 1/2 only the odd partials will be heard—and of these the fundamental will be considerably stronger than any of the others. This strength of the fundamental together with the absence of all even partial tones gives the resulting sound a somewhat 'hollow' tone.

(a) One method of verifying these statements is as follows:—

(1) Tune the 2nd string to make a unison with the 1st (see I (f) above).

(2) Pluck the 1st string alternately and accurately at 1/2 and 1/3 and note the difference in the tone quality. This results largely from the fact that, among other things, a fairly strong 3rd but no 2nd partial is heard when the string is plucked at 1/2, whereas, when plucked at 1/3 a fairly strong 2nd but no 3rd partial tone is heard. Also, the alternate sounding of the strong fundamental in the one case and the fairly strong 2nd and 4th partials in the other tends to give the impression of an octave jump. The absence of a particular partial is often easier to detect than its presence under these circumstances.

(3) If found necessary, the ear's attention can be easily directed to the partial tone in question by first striking the 2nd string, after having placed on it a node-locator at 1/2, 1/3, etc., and then listening for the corresponding partial in the complex tone from the 1st string.

Alternatively, verification of the statement made at the beginning of this section may be carried out using the principle of resonance (sympathetic vibration). In this method the 1st and 2nd strings are first tuned to unison as before. When this is done then it will be found that any vibrations occurring in the one string will cause corresponding, but weaker, vibrations in the other—the most important point being that the *only* vibrations occurring in the second string will be those contained in the vibration of the first. In other words, the 2nd string of the monochord can be used to detect the presence or absence of any particular vibration in the 1st string. Here, a node-locator on the 2nd string is particularly useful, since, placing it at 1/n will enable this string to respond only if there is sufficient of the n-th partial vibration present in the general vibration of the 1st string. The tuning of the unison must be good. Unless this is sufficiently accurate there will be no response.

(b) Applying now this method to the particular case in point, if the 1st string is plucked accurately at 1/2, and the node-locator placed, in turn on the 2nd string at 1/2, 1/3, 1/4, 1/5, etc., response will be found only to occur in the 2nd string when the locator is at 1/3, 1/5, etc., whereas, if the 1st string is plucked at 1/3, a fairly strong 2nd partial will be detected on the 2nd string, but no 3rd (or 6th etc.).

In order to be able to hear the response of the 2nd string it is necessary to damp out the vibrations of the 1st string almost immediately after plucking it. This is most conveniently done by touching it with the finger at about a centimetre from the point at which it was plucked.

The above technique can also be used, of course, on any stringed instrument—such as the piano, violin, guitar, etc. In this case, if a second of the instrument's strings is available, having any partials with frequencies equal to those for which it is desired to test in the first string, then a node-locator is not necessarily required. For example, if the interval between two strings is known to be a perfect fifth, then the frequency of the second partial vibration of the higher string will equal the frequency of the third partial vibration of the lower string, and the generation of one such partial vibration in either string will cause that partial having the same frequency in the other string to vibrate in sympathy with it.

III

Object. To show that the interval perceived by sounding the notes produced from two lengths of a stretched string, subjected to a constant tension, depends solely on the *ratio* of these lengths.

Also, to verify, *aurally*, that any such string-lengths having the particular ratios of $\frac{1}{2}$, $\frac{2}{3}$, $\frac{3}{4}$, $\frac{3}{5}$, $\frac{4}{5}$, $\frac{5}{6}$, $\frac{5}{8}$, etc., do in fact produce in the ear the consonant intervals of the octave, perfect fifth, perfect fourth, major sixth, major third, minor third, and minor sixth, etc., respectively.

Assuming the above to have been verified, it is easy to see that these ratios may be conveniently used to define uniquely (or to standardise), from their physical aspect, the intervals to which they correspond.

(a)

(1) Stop the string at $^1/_2$ and compare the note obtained by plucking this length with that obtained by plucking the open string.

(2) Stop the string at one-half of its length at (1), *i.e.* at $^1/_4$ [or $\frac{1}{2} \times \frac{1}{2}$, or $(\frac{1}{2})^2$] and compare its note with that obtained at $^1/_2$, and with the open string.

(3) Repeat as at (2) but with half that length, *i.e.* at $^1/_8$ [or $\frac{1}{2} \times \frac{1}{4}$, or $\frac{1}{2} \times \frac{1}{2} \times \frac{1}{2}$, or $(\frac{1}{2})^3$], and compare as before with $^1/_4$, $^1/_2$ and the open string (*i.e.* 1 or $^1/_1$).

(b)

(1) Next stop the string at $^2/_3$ and compare its note with that of the open string.

(2) Repeat (1) but take two-thirds of that length, *i.e.* stop at $^4/_9$ [or $\frac{2}{3} \times \frac{2}{3}$, or $(\frac{2}{3})^2$], and compare with the notes from $^2/_3$ and the open string.

(3) Repeat as at (2) but take two-thirds of that length, *i.e.* stop at $^8/_{27}$ [or $\frac{2}{3} \times \frac{4}{9}$, or $\frac{2}{3} \times \frac{2}{3} \times \frac{2}{3}$, or $(\frac{2}{3})^3$], and compare with $^4/_9$, $^2/_3$ and the open string.

(c) Repeat the whole of the procedure, as at (b), but with any other recognisable string-length ratio such as $\frac{3}{4}$, $\frac{3}{5}$, $\frac{4}{5}$, $\frac{5}{6}$, $\frac{5}{8}$ etc.

(d) Finally, take the note produced by any convenient length of the string as a starting or reference note (thereby automatically making its length 1, or its string-length ratio $\frac{1}{1}$) and compare this note with that obtained by sounding lengths of, say,

(1) $\frac{3}{5}$ of this length.

(2) $\frac{4}{5}$ of this length, etc.

(e) All the previous tests in this set have been melodic, *i.e.*, the notes have been sounded one after the other. But the most important thing about these intervals in Western music is their consonance—the ability of the ear to find or recognise them naturally and accurately when notes forming them are sounded simultaneously. In order to confirm this the following test should be tried out with two strings, sounding them simultaneously—or what is sometimes found to be more effective—in rapid succession.

(1) Tune the 2nd string to make a unison with the 1st, as in I (e).

(2) Stop, on the 1st string, any of the fractional lengths $\frac{1}{2}$, $\frac{2}{3}$, $\frac{3}{4}$, etc., sound the two strings and listen.

Best of all, see how near it is possible to get to these lengths, without looking at the calibrated scale—*i.e.* by ear alone.

After having achieved the object stated, it can be realised that these ratios provide a convenient and unique means of defining, or standardising—physically—the intervals to which they correspond.

In all this work it is important to realise that intervals cannot exist in the air. They have no physical existence whatever. They are effects on the aural perceptions—relations recognised only through the ear by the brain. Hence the importance of the word 'aurally' in the statement of the Object of the test.

Also, since consonant intervals are essentially harmonic, *i.e.* result from notes sounded simultaneously, it follows that the positions of the notes forming them are more or less clearly defined in the ear by the effects of the increased physical beating which occurs if they are mistuned. Such cannot be the case with purely melodic intervals, so that the notes forming them are thus naturally more fluid—depend more upon feeling and musical context—and do not therefore lend themselves so readily to a fixed physical definition. For instance, true major thirds (ratio $\frac{4}{5}$) and sixths (ratio $\frac{3}{5}$) when taken upwards in melody generally sound too flat and are naturally sharpened if the intonation is free, whereas, when taken in harmony there is no doubt that the true intervals sound the most pleasing. Intonation, if allowed to develop freely, is naturally flexible.

IV

Object. To show that the resultant string-length ratio obtained by multiplying together the string-length ratios corresponding to any given intervals is that ratio corresponding to the interval which is the sum of the two given intervals.

In short, that any two intervals having a note in common may be effectively *added* by *multiplying* their corresponding string-length ratios, and that they may be *subtracted* by *dividing* their corresponding string-length ratios, *i.e.* by *inverting* the ratio (or ratios) to be subtracted and then *multiplying* them together as for addition of intervals.

It is possible to deduce these facts directly from the previous test but it is perhaps better to verify them separately.

(a) It is a fact of aural experience when what the ear calls a fourth is added to what the ear calls a fifth the resultant interval is recognised by the ear as what it calls an octave.

Ignoring for the moment the supreme authority of the ear in these matters and accepting the physical definitions of these intervals as string-length ratios, then it is possible, in effect, to carry out this addition of intervals by calculation alone, as follows:—

(1) Choose any convenient string-length on the monochord.
(2) Next calculate the value of $^2/_3$ of this length.
(3) Now, calculate the value of $^3/_4$ of the length calculated at (2) above.
(4) Lastly, calculate the value of the ratio of the initial string-length, at (1), to the final string-length at (3).

This should come to $^1/_2$, which is not really surprising since $^2/_3 \times {}^3/_4 = {}^2/_4$ or $^1/_2$.

Thus, in terms of intervals:

5th + 4th = 8th (or octave)

or, in ratios $\frac{2}{3} \times \frac{3}{4} = \frac{1}{2}$

which can be checked by reference to the ratios listed in the previous test.

Note that we have not really managed without the ear, since the ratio $\frac{1}{2}$ only corresponds to an octave because the ear says so.

If considered necessary, this test should now be repeated as a practical aural experiment.

(b) In order to verify the case more generally the following pairs of consonant intervals should also be effectively added by calculation, *i.e.* using their corresponding ratios, and the results verified in the same manner as indicated at (a)

*m*3 + *M*3 = 5th

or ratios $\frac{5}{6} \times \frac{4}{5} = \frac{2}{3}$

*m*3 + 4th = *m*6

or ratios $\frac{5}{6} \times \frac{3}{4} = \frac{5}{8}$

*m*3 + *M*6 = 8th

or ratios $\frac{5}{6} \times \frac{3}{5} = \frac{1}{2}$

$$M3 + 4\text{th} = M6$$
$$\text{or ratios } \tfrac{4}{5} \times \tfrac{3}{4} = \tfrac{3}{5}$$
$$M3 + m6 = 8\text{th}$$
$$\text{or ratios } \tfrac{4}{5} \times \tfrac{5}{8} = \tfrac{1}{2}$$

(*c*) To verify the case for the subtraction of intervals (a) and/or (b) may be carried out in reverse, *e.g.*,

$$8\text{th} - 4\text{th} = 5\text{th}$$

(1) Choose any convenient length on the monochord.
(2) Next calculate the value of $^1/_2$ of this length.
(3) Now calculate the value of $^4/_3$ of the length calculated at (2)—(not $^3/_4$ since this interval is to be subtracted).
(4) Lastly, calculate the value of the ratio of the initial string-length, at (1) to the final string-length at (3).

The result should be $^2/_3$, since

$$8\text{th} - 4\text{th} = 5\text{th}$$
$$\text{or ratios } \tfrac{1}{2} \times \tfrac{4}{3} = \tfrac{2}{3}$$

For the sake of simplicity the number of intervals involved in the addition or subtraction in the above examples has been limited to two. This need not be so. For example,

$$m3 + M3 + 4\text{th} = 8\text{th}$$
$$\text{since ratios } \tfrac{5}{6} \times \tfrac{4}{5} \times \tfrac{3}{4} = \tfrac{1}{2}$$
$$\text{and } 8\text{th} - 4\text{th} + m3 - 5\text{th} = m3$$
$$\text{since ratios } \tfrac{1}{2} \times \tfrac{4}{3} \times \tfrac{5}{6} \times \tfrac{3}{2} = \tfrac{5}{6}$$

Whatever the physical quantity 'estimated' by the ear when it 'registers' the pitch of a note or the 'width' of an interval in the brain, it is certainly not the string-length or the string-length ratio, although, from the previous tests carried out with the monochord it obviously must be something which is proportional to these quantities. It must also, of course, be something common to all musical note production whatever the source—vibrating string, vibrating air column, etc.

Now that we know, that a musical 'note', before it reaches the ear, takes the form of a complicated periodic vibration in the air, and are able by using more or less elaborate equipment to examine and measure the physical properties of such a vibration, the previous remarks may seem superfluous. It is so easy nowadays to read and talk about the frequency, amplitude and harmonic content of such a vibration and of how these physical quantities affect the note heard, but this is by no means obvious and has taken many centuries and much careful thought to establish. Hence the importance of the monochord as a simple means of acquiring *first-hand* experience and stimulating interest in the fundamentals of musical acoustics.

Whereas the note heard depends on what the ear can make of the frequency, amplitude and harmonic content of a vibration, the interval perceived is found to depend almost entirely on the frequency-ratio of the vibrations involved—particularly when the notes are sounded simultaneously. But see also Chapter 13.

The measurement of audio frequencies involves the timing of vibrations occurring at a rate which is too rapid to be measured by simple methods and cannot be measured *directly* as such by means of a monochord. However, fortunately single calibrated standards of medium audio frequency are readily available in the form of tuning forks. Assuming such a fork of suitable frequency to be available, then the frequency of the monochord string can be easily and accurately obtained by the method of beating—as was done previously in Test I, section (e)—and hence the frequency of any other sounding length obtained from the ratio of these lengths, as shown in the Test below.

V

Object. To show that the frequency ratio corresponding to any given interval equals the inverse (or reciprocal) of the string-length ratio corresponding to the same interval.

This test requires a minimum of two forks of known frequency, for example, an 'a'' (440 cycles per second) and a 'c''' (523·2 cycles per second), *i.e.*, an equally-tempered minor third above the 'a''.

Ideally, a chromatic set of forks would be just the thing for this test.

(1) Assuming the open string of the monochord to have been tuned to 'a' (220 cyles per second) as explained in section 1 (e), then stopping the string at 500 mm, or at $^1/_2$ of its length, and plucking this length, should give a note making a perfect unison with the 'a'' fork.

(2) Now with the string still tuned to 'a', find and record the length required to produce a unison with the 'c″' fork.

Assuming the statement made in the Object to be so, then,

$$\frac{\text{length}_2}{\text{length}_1} \text{ should equal } \frac{\text{frequency}_1}{\text{frequency}_2}$$

$$\text{or length}_2 \text{ should equal } \frac{\text{frequency}_1}{\text{frequency}_2} \times \text{length}_1 =$$

$$\frac{440}{523{\cdot}2} \times 500 = 420{\cdot}5 \text{ mm}$$

If the test is carried out correctly then this length will be found to agree with that recorded at (2) above—within the practical limits of accuracy imposed by the method and apparatus used.

This fact is of great importance in the study of musical acoustics. Because of it 'notes' and 'intervals' obtained from a vibrating string can be correlated with those originating from any other type of musical instrument.

From the physical point of view the monochord provides a means whereby vibrations having given frequencies and frequency ratios can be obtained by measuring lengths and calculating their ratios. However, this method is limited to the monochord and does require a constant tension to be maintained in the string, whereas, if the frequency can be measured by a direct method, then no such proviso is necessary with any instrument.

Note also that an *increase* in frequency causes a corresponding *increase* in pitch, but an *increase* in the length of a vibrating string or column of air brings about a *decrease* in the pitch of the note heard. For this reason, and for those already stated, frequency and frequency ratio are now used universally in preference to string-length and string-length ratio, and except for Tests III and IV of this chapter, are so used throughout this book. *Unless otherwise stated, 'ratio' on its own is always used as a shortened form of 'frequency ratio'.*

When combined with IV, this experiment shows that the resultant frequency ratio obtained by multiplying together (or dividing) the frequency ratios corresponding to any two intervals is that ratio corresponding to the interval which is judged by the ear to be the sum (or difference) of the two separate intervals.

Or briefly—at least from a physical point of view—intervals can be added or subtracted by multiplication or division of the frequency ratios defining them.

VI

OBJECT. To acquire the technique of tuning the true consonant (harmonic) intervals of (a) the octave, (b) the fifth, (c) the fourth, (d) the major sixth, (e) the major third, (f) the minor third and (g) the minor sixth, using the strings of the monochord as a training device.

The main feature of this experiment is the clarity with which the tuning technique can be demonstrated. From a purely musical point of view the results can be somewhat disappointing. For in order to produce an octave between two strings of the same length, material and cross-section, the tension in the lower string has to be adjusted to a quarter of that in the higher string (which is here assumed to be already subjected to its maximum tension) and the consequent slackness in it is detrimental to the quality and definition of its note. However, with smaller intervals the effect is correspondingly smaller, and in all cases it can be reduced by plucking the string at a point as near to one end as is found practicable—giving it the minimum amount of deflection in the process.

In normal circumstances the required difference in tuning would be accommodated by using a shorter and/or lighter string for the higher note, or alternatively, using a longer and/or heavier string for the lower note.

The procedure for tuning these intervals is essentially the same as that already used to tune a unison (see Experiment I (f), page 192) except that here two differently placed node-locators are required for each adjustment.

In the instructions given below 'Lower' and 'Upper' refer to the pitch of the note perceived from the open string; 'Node at 1/2, 1/3 etc.' means that the string is to be divided by a node-locator at 1/2, 1/3 etc. of its open length. In all tests a unison is required between the appropriate partial tones, starting with the conditions given at '(1)', before proceeding to those given at '(2)'.

After some practice with these tests they should be tried without the aid of the node-locators, or at least without one of them.

(a) The Octave	(1)	(2)
Lower	Node at 1/2	Node at 1/4
Upper	Open string	Node at 1/2
	(3)	
Lower	Node at 1/6	
Upper	Node at 1/3	
(b) The Fifth	(1)	(2)
Lower	Node at 1/3	Node at 1/6
Upper	Node at 1/2	Node at 1/4
(c) The Fourth	(1)	(2)
Lower	Node at 1/4	Node at 1/8
Upper	Node at 1/3	Node at 1/6
(d) The Major Sixth	(1)	
Lower	Node at 1/5	
Upper	Node at 1/3	
(e) The Major Third	(1)	
Lower	Node at 1/5	
Upper	Node at 1/4	
(f) The Minor Third	(1)	
Lower	Node at 1/6	
Upper	Node at 1/5	
(g) The Minor Sixth	(1)	
Lower	Node at 1/8	
Upper	Node at 1/5	

VII

OBJECT. To tune true consonant chords.

This experiment requires the fitting of a third string to the monochord.

The technique is essentially the same as in the previous experiment—the tuning being carried out with pairs of strings, starting with the highest note and then 'slackening down' the other two strings to suit.

It is interesting to note that when setting up this experiment for the first inversion of the minor triad—the chord of the 6/3—the lowest pair of unison partials which define the major sixth of the chord, and the lowest pairs of unison partials which define its fourth and major third, together form the lowest common unison of partials of the chord. This is a special case, and comes about because the orders of the partials concerned in the common unison—3, 4 and 5—have not a common factor between them.

This particular lowest common partial can be heard quite plainly from the 4th, 5th and 6th open strings of a guitar—especially if the ear's attention is first drawn to it by lightly touching any one of these strings at the appropriate point.

When dealing with the minor triad it is the 4th, 5th and 6th partials which form the lowest common partial of the chord. (See Fig. A, Chapter 12.) But notice here that 6 and 4 have a common factor 2, which means that the lowest partials defining the fifth of the chord are an octave below these—namely the 3rd and 2nd.

It is somewhat surprising to discover that to tune a major triad in this particular manner it would be necessary to go as far as the fifth pair of unison partials between the two strings forming the fifth—*i.e.* the 10th and 15th partials—before reaching a common unison with a partial of the remaining string—its 12th partial. (See Fig A, Chapter 12.) But of course such high partials cannot be sounded and isolated satisfactorily with this simple monochord, and even if present in reasonable strength, could contribute little, if anything, to the definition of the major triad. They would, for all practical purposes, be masked by the presence and beating of the many lower partials —generally of greater strength. The major triad's strength of definition lies in its superiority in number of *pairs* of unisons between lower partials—such as the '3rd and 2nd', '5th and 4th', '6th and 4th', etc. as shown also in Fig. A—and in the position of its difference tones, as explained on page 19, and in Appendix 8.

The lowest partial tones of the notes of any chord which form a common unison may be calculated as follows.

Taking, for example, the first inversion of the minor triad, then the frequencies of the fundamentals of the vibrations will be in the ratio,

$$1 \quad \frac{5}{4} \quad \frac{5}{3}$$

Dividing them each by 5 makes their numerators all equal to 1 without disturbing their ratios of one to another. This gives

$$\frac{1}{5} \quad \frac{1}{4} \quad \frac{1}{3}$$

indicating that in the time required by
the *fundamental* of the first vibration to execute 1/5 of its cycle,
the *fundamental* of the second will have executed 1/4 of its cycle,
and
the *fundamental* of the third will have executed 1/3 of its cycle.

Since the 3rd, 4th, and 5th partials of each string will vibrate 3, 4, and 5 times as fast as their corresponding fundamental vibrations, it follows that, in the same length of time as above,

the *5th partial* vibration of the first string,
the *4th* ,, ,, ,, ,, second string,
and the *3rd* ,, ,, ,, ,, third string,

will each have completed exactly one cycle, *i.e.* will form a perfect unison.

In the same way, the ratio 1, $\frac{6}{5}$, $\frac{3}{2}$ of the minor triad, when divided through by 6 gives the ratio $\frac{1}{6}$, $\frac{1}{5}$, $\frac{1}{4}$, indicating that, when correctly intoned, a unison will be formed by the 6th, 5th and 4th partials of the appropriate strings—and so on. (See Fig. A, Chapter 12.)

The calculations used above to obtain the lowest partial (unison) tones of any chord may also be carried out in reverse, *i.e.* the order of the partials required to form a unison can be chosen first and the corresponding interval ratios then be calculated from them.

As an example of this, suppose that a unison was required between the 15th, 12th and 10th partials, then the calculations would be as follows.

The required interval ratios would be:

$$\frac{1}{15} \quad \frac{1}{12} \quad \frac{1}{10}$$

Multiplying these through by 15 gives:

$$1 \quad \frac{5}{4} \quad \frac{3}{2}$$

or alternatively, multiplying them through by 60 gives:

$$4 \quad 5 \quad 6$$

These ratios should be recognised as being those of the major triad—see the numbers given at the bottom of Fig. B, Chapter 12.

This, again, can be shown quite clearly and easily by means of the Harmonic Slide-rule (see Fig. A, Chapter 12).

These tests represent only the bare minimum of what can and should be carried out.

Free choice of melodic intervals makes another interesting test. I find that an upward step of a minor tone (ratio $\frac{10}{9}$) is too small, while a major tone (ratio $\frac{9}{8}$) is too large. My ear prefers a mean-tone in these circumstances! Also, whereas my ear prefers both its fourths and fifths to be true (ratios $\frac{4}{3}$ and $\frac{3}{2}$) whether taken melodically or harmonically, this is definitely not so for the major third and major sixth. In this case, although my ear prefers them to be true (ratios $\frac{5}{4}$ and $\frac{5}{3}$) when taken harmonically, when taken melodically as a leap these intervals sound very flat—especially the sixth—and have to be considerably sharpened if they are to satisfy it. I believe that this is so with most people and at least partly accounts for Palestrina's avoidance of the leap of a major sixth in his melodic lines. Some people are of the opinion that this is simply the influence of equal temperament imposed by the keyboard instruments. Speaking for myself—I have had practically nothing to do with keyboard instruments and believe that this is just one of those many, many facts for which no satisfactory explanation has yet been found—and quite possibly never will be.

19

ESTIMATION OF THE MAGNITUDE, SUM AND DIFFERENCE OF INTERVALS IN TERMS OF THEIR CORRESPONDING FREQUENCY RATIOS, AND THEIR PHYSICAL MEASUREMENT IN TERMS OF LOGARITHMIC UNITS

Re-writing, in terms of frequency ratios, the facts obtained from the previous tests on string-length ratios, gives the following:—

(i) The interval perceived from two notes sounded simultaneously depends on the frequency ratio of the vibrations from which they originate.

(ii) Pairs of vibrations having frequency ratios of $\frac{2}{1}$, $\frac{3}{2}$, $\frac{4}{3}$, $\frac{5}{3}$, $\frac{5}{4}$, $\frac{6}{5}$, $\frac{8}{5}$, etc., produce in the ear the consonant intervals of the 8th, 5th, 4th, *M*6, *M*3, *m*3, *m*6, etc.

(iii) Any intervals, having a note in common, may be effectively *added* by *multiplying* their corresponding frequency ratios, and effectively *subtracted* by first *inverting* the frequency ratio (or ratios) corresponding to the interval (or intervals) to be *subtracted*, and then *multiplying* as in the case of addition. This latter process does, of course, amount to division.

The relations between the pairs of consonant intervals recognised by the ear and shown previously to be obtainable by calculation from their corresponding string-length ratios in Section IV (a) and (b) of Chapter 18, can, if considered necessary, be re-calculated in terms of frequency ratios—the results in terms of intervals, being, as one would expect, exactly the same as before. For example

In terms of frequency ratios,

$$\frac{3}{2} \times \frac{4}{3} = \frac{2}{1}$$

or, in terms of intervals,

$$\text{5th} + \text{4th} = \text{8th} \qquad \text{(a)}$$

In the same way it follows that

$$M6 + m3 = 8\text{th} \qquad \text{(b)}$$
$$M3 + m6 = 8\text{th} \qquad \text{(c)}$$
$$4\text{th} + M3 = 6\text{th} \qquad \text{(d)}$$
$$M3 + m3 = 5\text{th} \qquad \text{(e)}$$
$$m3 + 4\text{th} = m6 \qquad \text{(f)}$$

Whenever such combinations of truly-tuned consonant intervals are made available on an instrument of fixed intonation, then their differences form smaller non-consonant adjacent intervals or steps which may be defined aurally and physically as follows:—

in terms of intervals, $M6 - 5\text{th} = t$ (g)
or, in terms of ratios, $\frac{5}{3} \times \frac{2}{3} = \frac{10}{9}$

so that t, termed a minor tone, will correspond to a frequency ratio of $\frac{10}{9}$.

$$5\text{th} - 4\text{th} = T \qquad \text{(h)}$$
$$\frac{3}{2} \times \frac{3}{4} = \frac{9}{8}$$

so that T, termed a major tone, will correspond to a frequency ratio of $\frac{9}{8}$.

$$4\text{th} - M3 = s \qquad \text{(i)}$$
$$\frac{4}{3} \times \frac{4}{5} = \frac{16}{15}$$

so that s, termed a (diatonic) semitone, will correspond to a frequency ratio of $\frac{16}{15}$.

By suitably combining these equations with equations (a) to (f) it is possible to write down all the consonant intervals in terms of these steps—as listed below:—

$$m3 = T + s \qquad \text{(j)}$$
$$M3 = T + t \qquad \text{(k)}$$
$$4\text{th} = M3 + s = T + t + s \qquad \text{(l)}$$
$$5\text{th} = 4\text{th} + T = 2T + t + s \qquad \text{(m)}$$
$$m6 = 5\text{th} + s = 2T + t + 2s \qquad \text{(n)}$$
$$M6 = 5\text{th} + t = 2T + 2t + s \qquad \text{(o)}$$
$$8\text{th} = M6 + m3 = 3T + 2t + 2s \qquad \text{(p)}$$

and so on.

These relationships can be checked arithmetically from the frequency ratios of their intervals. They are indicated and displayed in Chapter 20 and Figs. C and D of Chapter 16.

Because of their association with the diatonic scale the small intervals defined by equations (g), (h) and (i) are usually known as diatonic steps. They are also known as secondary intervals because of the indirect way in which they are formed.

In any system employing free intonation, such steps could only be generated with any precision and consistency in a purely harmonic progression. (See, for example, Just Temperament, Chapter 15.) In melodic usage their width would differ considerably according to taste and context and could no longer be assumed to be defined physically in terms of these exact frequency ratios.

One advantage gained by expressing intervals in terms of these diatonic steps is that the '*t*'s', '*T*'s' and '*s*'s' need only to be added or subtracted in order to add or subtract the intervals they form—though if in any particular case the resultant combination of these steps cannot be recognised as a diatonic interval then it becomes necessary to revert to the method of calculation by multiplying interval ratios.

In order to demonstrate the two methods so far mentioned consider the following problem.

Which is the greater interval, four perfect fifths or two octaves and a major third? Express the difference as an interval ratio.

(a) By calculation using interval ratios.

$$4(5\text{th}) = \text{ratio } (\tfrac{3}{2})^4 \text{ or } \tfrac{3}{2} \times \tfrac{3}{2} \times \tfrac{3}{2} \times \tfrac{3}{2} = \tfrac{81}{16}$$

$$2(8\text{th}) + M3 = \text{ratios } \tfrac{2}{1} \times \tfrac{2}{1} \times \tfrac{5}{4} = \tfrac{20}{4} = \tfrac{20 \times 4}{4 \times 4} = \tfrac{80}{16}$$

Since $\frac{81}{16}$ is obviously greater than $\frac{80}{16}$ it follows that four fifths span a greater interval than do two octaves and a major third.

To obtain the ratio of the interval by which $\frac{81}{16}$ exceeds $\frac{80}{16}$ the former must be divided by the latter, *i.e.*,

$$\tfrac{81}{16} \times \tfrac{16}{80} = \tfrac{81}{80}$$

(b) By diatonic steps—followed by calculation making use of equations (m), (p) and (k),

$$4(5\text{th}) = 4(2T + t + s) = 8T + 4t + 4s$$

$$2(8\text{th}) + M3 = 2(3T + 2t + 2s) + (T + t) = 7T + 5t + 4s$$

Here it is not so obvious which of these intervals is the greater, though as in this particular case both intervals contain four semitones, the problem can be reduced to finding which is the greater—

$$8T + 4t \text{ or } 7T + 5t$$

This can, in turn, be further simplified by subtracting $7T + 4t$ from both intervals. The problem is then finally reduced to finding which is the greater,

$$T \text{ or } t$$

Reverting now to interval ratios, since $T = \frac{9}{8}$ (*i.e.* 1 and $\frac{1}{8}$) and $t = \frac{10}{9}$ (*i.e.* 1 and $\frac{1}{9}$)—and $\frac{1}{8}$ is greater than $\frac{1}{9}$—it follows that four perfect fifths are greater than two octaves and a major third—the exact amount being:—

$$T - t = \text{ratio } \frac{9}{8} \times \frac{9}{10} = \frac{81}{80}$$

This interval $T - t$, or ratio $\frac{81}{80}$ (denoted by k) becomes important in the study of temperaments and is known as the comma of Didymus. It can easily be sounded by means of the monochord.

In the case just considered it is fairly obvious to most people that $\frac{9}{8}$ is greater than $\frac{10}{9}$, but this cannot be said of, say, $\frac{32}{25}$ and $\frac{41}{32}$. The following are examples of the many ways of finding the greater of two interval ratios, and later, of measuring this difference. For the sake of simplicity these methods are first demonstrated on the ratios $\frac{9}{8}$ and $\frac{10}{9}$.

(c1) By expressing them with common denominators

$$\frac{9}{8} = \frac{9 \times 9}{8 \times 9} = \frac{81}{72}$$

$$\frac{10}{9} = \frac{10 \times 8}{9 \times 8} = \frac{80}{72}$$

since $\frac{81}{72}$ is greater than $\frac{80}{72}$
therefore $\frac{9}{8}$ is greater than $\frac{10}{9}$.

(c2) By expressing them with common numerators

$$\frac{9}{8} = \frac{9 \times 10}{8 \times 10} = \frac{90}{80}$$

$$\frac{10}{9} = \frac{10 \times 9}{9 \times 9} = \frac{90}{81}$$

since $\frac{90}{80}$ is greater than $\frac{90}{81}$
therefore $\frac{9}{8}$ is greater than $\frac{10}{9}$.

(c3) In this particular case the following interesting property of fractions may be used to show which is the greater ratio.

If both the numerator and denominator of a fraction—or ratio expressed as a fraction—are increased by the same amount, then the fraction is moved nearer to 1 (or $\frac{1}{1}$)

e.g. $\frac{3+2}{4+2} = \frac{5}{6}$ which is nearer to 1 than is $\frac{3}{4}$

and $\frac{5+2}{4+2} = \frac{7}{6}$ which is nearer to 1 than is $\frac{5}{4}$.

In the same way, if the same amount is subtracted, then the fraction or ratio is moved further away from 1 or unity.

Applying now this principle to the interval ratios of $\frac{9}{8}$ and $\frac{10}{9}$, then since $\frac{9+1}{8+1} = \frac{10}{9}$, then $\frac{10}{9}$ must be nearer to 1 than to $\frac{9}{8}$

or alternatively,

since $\frac{10-1}{9-1} = \frac{9}{8}$, then $\frac{9}{8}$ must be further from 1 than $\frac{10}{9}$

hence, in both cases, $\frac{9}{8}$ is greater than $\frac{10}{9}$.

(c4) By expressing them as decimal fractions

$$\frac{9}{8} = 9 \div 8 = 8\overline{)\,9{\cdot}000} = 1{\cdot}125$$
$$1{\cdot}125$$

$$\frac{10}{9} = 10 \div 9 = 9\overline{)10{\cdot}000} = 1{\cdot}111$$
$$1{\cdot}111 \text{ etc.}$$

hence, as before, $\frac{9}{8}$ is greater than $\frac{10}{9}$.

(d) In the methods used above at (c), each ratio is expressed in such a form that it is easy to see, by mere inspection, that there is a numerical *difference* between them and hence, which of the two ratios is numerically the greater. It is not necessary, for this purpose, to work out the actual numerical difference—and even if this were done, the result would not help in any way in determining the corresponding interval by which the ratios differed.

Another way of comparing two ratios is to find how many *times* one is smaller (or larger) than the other, or is contained in the other, *i.e.* to find the *ratio* of the two ratios, by *division*. This would appear to be a more natural way of doing the job since it expresses the *difference* between the two ratios *in terms of the same sort of thing—another ratio*. It has, in fact, already been used at the end of the previous sections (a) and (b) and has the advantage of giving both a comparison of the magnitude of ratios, and a statement, in the form of a ratio, from which the interval corresponding to their difference can be physically *defined* or *standardised.*

Applying this method to frequency ratios of $\frac{32}{25}$ and $\frac{41}{32}$

$$\text{then } \frac{\frac{41}{32}}{\frac{32}{25}} = \frac{41}{32} \div \frac{32}{25} = \frac{41}{32} \times \frac{25}{32} = \frac{1025}{1024}$$

Hence the interval corresponding to the ratio $\frac{41}{32}$ is greater than the interval corresponding to the ratio $\frac{32}{25}$ by the interval corresponding to the ratio $\frac{1025}{1024}$.

(e) By the conversion of frequency ratios to logarithmic units—as shown in the following section at (b). See also *standardisation and measurement of intervals* in Chapter 12.

This is the best method of all, since its use enables *any* ratio—and therefore its corresponding physical interval—to be not only compared and defined but also measured.

First, however, consider the reverse of the processes discussed so far.

Whenever intervals are added aurally—say, a fifth + a fourth—then, in a manner of speaking, the ear, in saying that the sum is an octave, may be considered, in effect, to have multiplied the ratios corresponding to these intervals—although it would appear to know nothing of the mathematical process normally required to carry out this operation.

In a similar manner, whenever the indices of two powers of the same number are added together *e.g.*, the '1' and '2' of, say, 2^1 and 2^2, the result, 2^3, equals the product of 2^1 and 2^2 (or $2 \times 4 = 8$) *i.e.* the product of the powers. In terms of intervals this result can be translated as meaning that 1 octave + 2 octaves = 3 octaves, so that to express interval ratios in terms of powers of 2 is in effect, to express them in terms of octaves. This is alright so far as octaves are concerned but can intervals lying between octaves be expressed in this system? What can be safely said without fear of contradiction is that powers of 2 corresponding to the ratios of these intermediate intervals must include fractions (vulgar or decimal), and that for intervals lying between a unison and an octave the corresponding powers of 2 must lie between 0 and 1 ($2^0 = 1$, see Chapter 21), for intervals between 1 and 2 octaves, powers between 1 and 2, etc. Fortunately, cutting a long story short, all these fractional powers have been worked out and are readily available in the form of logarithms, or in specially adapted multiples of them known as cents, savarts, etc.

A table of cents (and savarts) at 1 cent intervals together with the corresponding interval ratios—expressed in decimal form—is available at the end of the book. With the aid of this table it is possible to obtain a *numerical measure* (to the nearest cent) of any interval ratio which agrees more or less accurately with that of the ear's estimation of the interval from the corresponding vibrations—and vice-versa.

(The figures given below in square brackets are more exact versions of those that come immediately before them.)

There are 1200 cents to the octave, *i.e.*, 100 for each semitone of equal temperament.

1 cent corresponds to a ratio of about

$$\frac{1731}{1730} \text{ or } 1 + \frac{1}{1730} = 1{\cdot}0006\ [1{\cdot}000576]$$

There are 301 savarts to the octave, *i.e.*, about 25 [25·083] for each semitone of equal temperament.

1 savart corresponds to a ratio of about

$$\frac{436}{435} \text{ or } 1 + \frac{1}{435} = 1{\cdot}002\ [1{\cdot}002305]$$

1 savart = 4 [3·986] cents.

1 cent = $^1/_4$ or ·25 [·25086] savarts.

For more details of these logarithmic units see Chapters 20 and 21.

The Calculation of Ratios in Terms of Cents and Cents in Terms of Ratios Using the Ratio Table

(a) To obtain the ratio corresponding to a given number of cents.

Taking, as a first example, a semitone of equal temperament, then,

$$1 \text{ semitone} = \frac{1200}{12} = 100 \text{ cents}$$

From the third column of the '100' cents line of the Ratio Table it can be seen that the required ratio, in its decimal form, is '1·0595', or 1·0595/1.

If this decimal ratio is required to be expressed in the form of a vulgar fraction, then it can be obtained—as can any other decimal fraction—exactly, by the method shown in Chapter 21, or to a suitable degree of approximation, by the method given in Appendix 9.

As a second example, consider the 'fifth' of Robert Smith's temperament, in which,

$$1 \text{ 'fifth'} = 1200 \times \tfrac{29}{24} = 696 \text{ cents}$$

From the third column of the '696' cents line of the Ratio Table it can be seen that the required ratio is '1·4948', or 1·4948/1.

Next consider the 'fifth' of the cycle of 31. In this temperament,

$$1 \text{ 'fifth'} = 1200 \times \tfrac{18}{31} = 696{\cdot}774 \text{ cents}$$

To the nearest cent this is 697, for which the Ratio Table gives 1·4957.

However, a slightly more accurate figure can be found by taking the fractional part of this value into account, as follows,

From the Ratio Table 697 cents = 1·4957
and 696 cents = 1·4948
therefore, subtracting, 1 cent = ·0009 change at this ratio level
and ·77 cent = ·0009 × ·77
= ·000693, say, ·0007
therefore 696 + ·77 cents = 1·4948 + ·0007
or 696·77 cents = 1.4955

Finally, consider the 'fifth' of Silbermann's temperament, which equals a true fifth minus one-sixth of a comma,

$$\begin{aligned} 1 \text{ 'fifth'} &= 701{\cdot}955 - \tfrac{21{\cdot}506}{6} \text{ cents} \\ &= 701{\cdot}955 - 3{\cdot}584 \text{ cents} \\ &= 698{\cdot}371, \text{ say, } 698{\cdot}37 \text{ cents} \end{aligned}$$

To the nearest cent this is 698, for which the Ratio Table gives 1·4966.

But, as previously, a slightly more accurate figure, if thought necessary, can be obtained by taking the '·37' into account.

From the Ratio Table 699 cents = 1·4974
and 698 cents = 1·4966

therefore, subtracting, $\underline{1 \text{ cent} = \cdot 0008}$ change at this ratio level

and $\cdot 37 \text{ cent} = \cdot 0008 \times \cdot 37$

$= \cdot 000296$, say $\cdot 0003$

$698 + \cdot 37 \text{ cents} = 1 \cdot 4966 + \cdot 0003$

or $698 \cdot 37 \text{ cents} = 1 \cdot 4969$

Independent checks can often be made on certain ratios occurring in a regular temperament. In Basic Mean-tone, for instance, since all major thirds are kept true, it follows that all notes a major third apart must have corresponding frequency ratios of $^5/_4$. Similarly, in Silbermann's temperament, since each tempered fifth is flattened by $^1/_6$ of a comma, it follows that any six of these tempered fifths must equal six true fifths minus a comma. So that in this temperament, for example:—

interval c — f♯''' $= (\frac{3}{2})^6 \times \frac{80}{81} = \frac{729}{64} \times \frac{80}{81} = \frac{45}{4}$

∴ interval c — f♯ $= \frac{45}{4} \times \frac{1}{8} = \frac{45}{32}$ *exactly*

again, in the same way,

interval c — G♭,,, $= (\frac{2}{3})^6 \times \frac{81}{80} = \frac{64}{729} \times \frac{81}{80} = \frac{4}{45}$

∴ interval c — g♭ $= \frac{4}{45} \times \frac{16}{1} = \frac{64}{45}$ *exactly*

and so on.

(b) To obtain the number of cents corresponding to a given ratio.

If not already in decimal form, the given ratio must be so expressed before its corresponding value in cents can be obtained from the Ratio Table. To do this the numerator ('top') must be divided by the denominator ('bottom'). Alternatively, the numerator may be multiplied by the reciprocal of the denominator, *i.e.* by 1/the denominator. A table of the reciprocals of numbers, from 1 to 100, is given for this purpose at the end of the book.

Sometimes it is required to find the ratio corresponding to a given number of cents (or savarts). This requires the tables to be used in the other direction—or alternatively—the formula given for *r* in Chapter 20 may be used.

Reverting now to the simple problem used for demonstrating the previous methods:—

$\frac{9}{8}$ or $1 + \frac{1}{8} = 1 + \cdot 125 = 1 \cdot 125$

The nearest decimal fraction to this listed in the table is

1·1251, the corresponding value in cents being 204. This is near enough for most purposes.

$$\tfrac{10}{9} \text{ or } 1 + \tfrac{1}{9} = 1 + \cdot 1111 = 1\cdot 1111$$

This value lies somewhat nearer to 1·1109 than to 1·1115, so that its value in cents lies somewhat nearer to 182 than to 183. Again, 182 is accurate enough for most practical purposes. Therefore $\frac{9}{8}$ is greater than $\frac{10}{9}$ by 204 − 182 = 22 cents.

From what has been done previously at (a) and (b) of the previous section, this answer should be equivalent to the ratio $\frac{81}{80}$. Now

$$\tfrac{81}{80} \text{ or } 1 + \tfrac{1}{80} = 1 + \cdot 0125 = 1\cdot 0125.$$

This decimal fraction lies exactly halfway between 1·0122 and 1·0128 in the table so that its corresponding value in cents lies about halfway between 21 and 22 cents, *i.e.* 21·5 cents. This compares very favourably with the more exact figure of 21·506.

Generally, as above, if the decimal fraction whose corresponding value in cents is required does not occur exactly in the table, then a slightly more accurate figure may be obtained by splitting, in direct proportion, the difference between the two nearest values—and vice-versa when obtaining the value of a ratio from its value in cents.

As further examples of this technique consider again the ratios $\frac{9}{8}$ and $\frac{10}{9}$, neither of which corresponded exactly with any decimal fraction given in the table, *i.e.*

$\frac{9}{8}$ = 1·125 whereas, from the table	1·1251 = 204 cents
	1·1244 = 203 cents
subtracting these	·0007 = 1 cent

To make 1·1244 up to 1·125 requires an additional ·0006,
since ·0007 corresponds to 1 cent—at this ratio level
then ·0001 „ to about $\frac{1}{7}$ cent
and ·0006 „ to about $\frac{6}{7}$ cent

From the table of reciprocals $^1/_7$ = ·14286, but ·143 will be near enough for the purpose. So that

$\frac{9}{8} = 203 + \frac{6}{7} = 203 + 6 \times \cdot 143 = 203 + \cdot 858$ which equals 203·9 to four significant figures. This compares very well with the more exact figure of 203·910 cents.

In like manner

$\frac{10}{9} = 1{\cdot}1111$ whereas, from the table $\begin{array}{l} 1{\cdot}1115 = 183 \text{ cents} \\ 1{\cdot}1109 = 182 \text{ cents} \end{array}$

subtracting these $\quad {\cdot}0006 = \quad 1 \text{ cent}$

To make 1·1109 up to 1·1111 requires an additional ·0002.
since ·0006 corresponds to 1 cent—at this ratio level
then ·0001 ,, to about $\frac{1}{6}$ cent
and ·0002 ,, to about $\frac{1}{3}$ cent

From the table of reciprocals ⅓ = ·33333

So that

$\frac{10}{9} = 182 + {\cdot}3333 = 182{\cdot}3$ to four significant figures. A more exact figure is 182·404 cents.

Using the figures just obtained from the tables, as before $\frac{9}{8}$ is greater than $\frac{10}{9}$ by 203·9 — 182·3 = 21·6 cents—which is nearer to the more exact figure of 21·506 than is the 22 cents previously obtained when working with the table to the nearest cent.

Below is given a list of intervals and their corresponding ratios expressed to the nearest cent, from the table given—followed by a more exact figure in square brackets.

$k=\frac{81}{80}$	$=1+\frac{1}{80}$	$=1+{\cdot}0125$	$=1{\cdot}0125$	$=22$ cents	[21·506]
$s=\frac{16}{15}$	$=1+\frac{1}{15}$	$=1+{\cdot}0667$	$=1{\cdot}0667$	$=112$ cents	[111·731]
$t=\frac{10}{9}$	$=1+\frac{1}{9}$	$=1+{\cdot}1111$	$=1{\cdot}1111$	$=182$ cents	[182·404]
$T=\frac{9}{8}$	$=1+\frac{1}{8}$	$=1+{\cdot}125$	$=1{\cdot}125$	$=204$ cents	[203·910]
$m3=\frac{6}{5}$	$=1+\frac{1}{5}$	$=1+{\cdot}2$	$=1{\cdot}2$	$=316$ cents	[315·641]
$M3=\frac{5}{4}$	$=1+\frac{1}{4}$	$=1+{\cdot}25$	$=1{\cdot}25$	$=386$ cents	[386·314]
4th$=\frac{4}{3}$	$=1+\frac{1}{3}$	$=1+{\cdot}3333$	$=1{\cdot}3333$	$=498$ cents	[498·045]
5th$=\frac{3}{2}$	$=1+\frac{1}{2}$	$=1+{\cdot}5$	$=1{\cdot}5$	$=702$ cents	[701·955]
$m6=\frac{8}{5}$	$=1+\frac{3}{5}$	$=1+{\cdot}6$	$=1{\cdot}6$	$=814$ cents	[813·687]
$M6=\frac{5}{3}$	$=1+\frac{2}{3}$	$=1+{\cdot}6667$	$=1{\cdot}6667$	$=884$ cents	[884·359]
8th$=\frac{2}{1}$	$=2$		$=2$	$=1200$ cents	[1200·00]

Using these figures to check the relations between the consonant intervals already obtained from their ratios, gives the following:—

$$m3 + M3 = 316 + 386 = 702 = 5\text{th}$$
$$m3 + 4\text{th} = 316 + 498 = 814 = m6$$
$$M3 + 4\text{th} = 386 + 498 = 884 = M6$$
$$M3 + m6 = 386 + 814 = 1200 = 8\text{th}$$
$$4\text{th} + 5\text{th} = 498 + 702 = 1200 = 8\text{th}$$

At the beginning of Chapter 6, on pages 53 and 54, there are four equations which define four of those small intervals that occur as discrepancies whenever attempts are made to fit true fifths and true major thirds to octaves. While at the end of the same chapter the relative sizes of these discrepancies are simply stated as facts. The reader should now be in a position to calculate the ratios to which these intervals correspond and to verify the statements about their relative sizes, starting from the given equations, as follows.

In terms of ratios,
from eq. (1)

$$k = 4(5\text{th}) - 2(8\text{th}) - M3$$
$$= \frac{3^4}{2^4} \times \frac{1}{2^2} \times \frac{4}{5} = \frac{81}{80} \text{ (or } 1\cdot0125\text{) } \textit{exactly} \qquad \text{(i)}$$

from eq. (2)

$$s = 8(5\text{th}) + M3 - 5(8\text{th})$$
$$= \frac{3^8}{2^8} \times \frac{5}{4} \times \frac{1}{2^5}$$
$$= 3^8 \frac{\times 5}{2^{15}} = \frac{6561 \times 5}{32769}$$
$$= \frac{32805}{32768} \textit{ exactly} \text{ or } 1\cdot000113 \ldots \text{ approx.} \qquad \text{(ii)}$$

from eq. (3)

$$p = 12(5\text{th}) - 7(8\text{th})$$
$$= \frac{3^{12}}{2^{12}} \times \frac{1}{2^7} = \frac{3^{12}}{2^{19}}$$
$$= \frac{531441}{524288} \textit{ exactly} \text{ or } 1\cdot0136 \ldots \text{ approx.} \qquad \text{(iii)}$$

from eq. (4)

$$d = 8\text{th} - 3\,(M3)$$
$$= \frac{2}{1} \times \frac{4^3}{5^3} = \frac{2}{1} \times \frac{64}{125} = \frac{128}{125} \text{ or } 1\cdot024 \text{ exactly} \qquad \text{(iv)}$$

There is a list of powers of numbers in the tables at the end of the book.

And now repeating this procedure in logarithmic units, *i.e.*, measures or ratio, we have,

In terms of cents,

$$k = 4(5\text{th}) - 2(8\text{th}) - M3$$
$$= 4 \times 701\cdot955 - 2 \times 1200 - 386\cdot314$$

$$= 2807{\cdot}820 - 2400 - 386{\cdot}314$$
$$= 2807{\cdot}820 - 2786{\cdot}314$$
$$= 21{\cdot}506 \text{ cents} \qquad \text{(v)}$$
$$s = 8(5\text{th}) + M3 - 5(8\text{th})$$
$$= 8 \times 701{\cdot}955 + 386{\cdot}314 - 5 \times 1200$$
$$= 5615{\cdot}640 + 386{\cdot}314 - 6000$$
$$= 6001{\cdot}954 - 6000$$
$$= 1{\cdot}954 \text{ cents} \qquad \text{(vi)}$$
$$p = 12(5\text{th}) - 7(8\text{th})$$
$$= 12 \times 701{\cdot}955 - 7 \times 1200$$
$$= 8423{\cdot}460 - 8400$$
$$= 23{\cdot}460 \text{ cents} \qquad \text{(vii)}$$
$$d = 8\text{th} - 3(M3)$$
$$= 1200 - 3 \times 386{\cdot}3137$$
$$= 1200 - 1158{\cdot}941$$
$$= 41{\cdot}059 \text{ cents} \qquad \text{(viii)}$$

The statements made on page 58 concerning the approximate relative sizes of these small intervals can now be checked, in cents, as follows:—

$11s = 11 \times 1{\cdot}954 = 21{\cdot}494 =$ one comma (k)—very nearly.
$12s = 12 \times 1{\cdot}954 = 23{\cdot}448 =$ one comma (p)—very nearly.
$21s = 21 \times 1{\cdot}954 = 41{\cdot}034 =$ one diesis (d)—very nearly.

also,

$5d = 5 \times 41{\cdot}059 = 205{\cdot}30$ one major tone (T)—very nearly.

Equations (5) and (6), on page 54, are left to the reader to verify.

In order to represent intervals graphically in such a manner that they may be added or subtracted by simply adding or subtracting the length representing them, this length must be made proportional to the logarithm of their ratio—or to some multiple of this logarithm. Only by so doing is an interval represented by the same length whatever its position in pitch.

Finally, one good argument in favour of expressing (or defining) intervals physically in terms of the ratio of two whole numbers is that this is the only way that some of the consonant intervals and their combinations can be represented *exactly*. No normal person recognises the ratio $\frac{1{\cdot}6667}{1}$ as anything simple until or unless he 'translates' it into the ratio $\frac{5}{3}$—the latter being exactly what is meant but only approximately indicated by the former.

20

THE CALCULATION OF THE FREQUENCY RATIOS OF THE INTERVALS USED IN TUNINGS AND TEMPERAMENTS AND THE FREQUENCIES CORRESPONDING TO THE NOTES REQUIRED TO PRODUCE THEM

In determining the frequency ratio of an interval it is usual to divide the higher frequency by the lower, *i.e*,, to use the lower frequency as a yardstick with which to measure the higher. When this is done the interval is, in effect, regarded in an upward direction from the lower pitched note, and it follows that the resulting ratio must be greater than unity—or the 'top' must be greater than the 'bottom'. When so expressed, all intervals lying between the unison and octave will correspond with frequency ratios lying between $\frac{1}{1}$ (or 1) and $\frac{2}{1}$.

Alternatively, the lower frequency may be divided by the higher, *i.e.*, the higher frequency used as a yardstick or unit with which to measure the lower. In this case the interval is, in effect, regarded in a downward direction from the higher pitched note, and it follows that the resulting ratio must be less than unity (or '1').

If, in the manner indicated above, the frequency of a 'note' be made to correspond with 1, then the note a major third above it must correspond with a frequency of $\frac{5}{4}$, or 1 and $\frac{1}{4}$, and the note a major third below with a frequency of $\frac{4}{5}$. In other words, in the time taken for one cycle of the reference 'note' (that note corresponding to '1') to occur, there will occur 1 and $\frac{1}{4}$ cycles of the 'note' a major third above it and $\frac{4}{5}$ of a cycle of the 'note' a major third below it. Looked at in this way an interval measured downwards can be thought of as an (upwards) interval subtracted from the reference note, represented by '1', as follows:—

$$1 - \text{M}3 = 1 \div \frac{5}{4} = 1 \times \frac{1}{\frac{5}{4}} = 1 \times \frac{4}{5} = \text{M}3 \text{ measured downwards.}$$

Multipilying any interval ratio by 2, 4, 8, etc., increases the interval it represents by 1, 2, 3, etc., octaves.

Conversely, dividing an interval ratio by 2, 4, 8, etc., decreases the interval it represents by 1, 2, 3, etc., octaves.

Just Temperament and the Pure Scale

It is proposed, merely as a starting point, to consider the theoretical tuning of a keyboard instrument with fixed intonation in which it is required to provide as a minimum harmonic resource, two conjunct fifths (F—C—G, for instance) with major triads formed on each of the notes defining these fifths; to accept as a matter of course that all intervals so formed are to be repeated exactly within each octave (see *standard form*, Chapter 12); and, if only for the sake of simplicity, to ignore at this stage the melodic demands of any 'unessential notes' which would contradict or at least modify this choice of intervals.

Next, to determine theoretically the intervals which each of the notes defined as above make with the note (C) that is common to the two fifths, with their adjacent notes, and with any other notes of this tuning.

Before starting on this scheme it is worth noting that, because of repetition at the octave, every interval introduced to such a tuning will be accompanied—in the opposite direction—by its inversion. For example, an upward fifth must introduce a downward fourth; a major sixth, a minor third; a major second, a minor seventh, etc.; and vice-versa.

Starting, purely for ease in calculation, with the note associated with a frequency of 240 cycles per second, and denoting it, for simplicity, by C, then the note c, one octave above it, will correspond with a frequency of $240 \times 2 = 480$ cycles per second.

In the same way the note a fifth above C—*i.e.*, G, the dominant of C—will correspond with a frequency of $240 \times \frac{3}{2} = 360$, while that of the fifth below c—*i.e.*, F, the subdominant of C—will correspond with a frequency of $480 \times \frac{2}{3} = 320$ cycles per second.

These notes may be written out as follows:—

C	F	G	c
240	320	360	480

Continuing . . . the frequency corresponding to the note a major third above C (*i.e.*, E) will be $240 \times \frac{5}{4} = 300$ cycles per second, that a major third above F (*i.e.*, A) with $320 \times \frac{5}{4} = 400$, and that a major third above G (*i.e.*, B) with $360 \times \frac{5}{4} = 450$ cycles per second. Finally, the frequency corresponding with the note a fifth above G (*i.e.*, d) will be $360 \times \frac{3}{2} = 540$ cycles per second.

The group of notes obtained above constitute what is known as *just temperament.* It follows from the method of their derivation that these notes must produce true intonation for the major triads C E G, F A c, and G B d,—and their inversions.

The frequency corresponding with the note D may be obtained by dividing the frequency corresponding with the note d by 2, thereby lowering the note by an octave. This makes D correspond with a frequency of $\frac{540}{2} = 270$ cycles per second.

Writing out, in order, a complete octave of these notes, together with the frequencies with which they correspond, gives the following:—

C	D	E	F	G	A	B	c	
240	270	300	320	360	400	450	480	.. (a)

Now, working backwards, in order to obtain the interval which each of these notes makes with the keynote, C must be made the reference note, *i.e.*, the frequencies corresponding to the notes of group (*a*) above must be divided by 240, the frequency corresponding with C.

When this is done, and each ratio so obtained expressed in its lowest terms, the result may be conveniently shown as below—the intervals to which the ratios correspond being given in brackets.

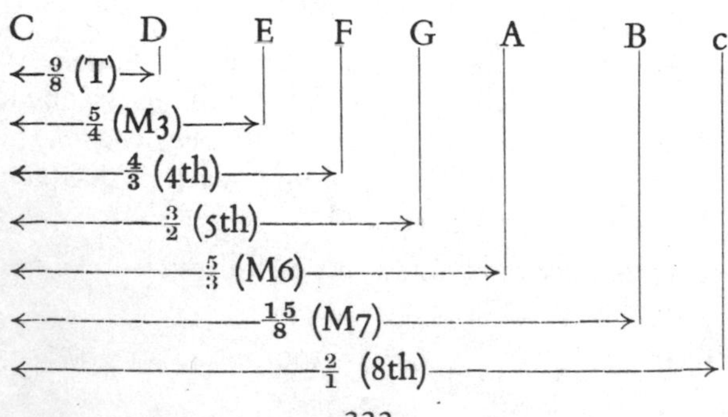

Once the (frequency) ratios corresponding to any given set of intervals are available, then it is a simple matter to calculate the frequencies of the vibrations required to reproduce the same set of intervals at any chosen level of pitch. All that is necessary is to multiply the appropriate ratios by that frequency associated with the pitch of the chosen reference note. This has been done below in the last line of figures for a reference note corresponding to a frequency of 264 cycles per second.

The frequency '264' has been chosen because—in this particular case—it gives to 'A' its arbitrary frequency of 440 (defined as corresponding to *standard musical pitch*). This can be easily shown as follows:

$$\text{since } \frac{\text{`C's' frequency}}{\text{`A's' frequency}} = \frac{\text{`C's' frequency}}{440} = \frac{3}{5}$$

it follows that 'C's' frequency must be equal to

$$440 \times \tfrac{3}{5} = 88 \times 3 = 264 \text{ cycles per second.}$$

Below, at (b), are listed the ratios defining the intervals of just temperament together with their corresponding values in cents, string-lengths, etc.

C	D	E	F	G	A	B	c	
1	$\frac{9}{8}$	$\frac{5}{4}$	$\frac{4}{3}$	$\frac{3}{2}$	$\frac{5}{3}$	$\frac{15}{8}$	$\frac{2}{1}$	. . . . (b)
0	203·9	386·3	498·0	702·0	884·4	1088·3	1200	cents
1000	888·8	800·0	750·0	666·7	600·0	533·3	500·0	string-lengths
264·0	297·0	330·0	352·0	396·0	440·0	495·0	528·0	frequencies for A–440

The diatonic steps between adjacent notes of this group can now be calculated from their corresponding ratios as follows:—

The interval between—

C and D=ratio $\frac{\frac{9}{8}}{1} = \frac{9}{8} = T = 203 \cdot 9$ cents

D and E=ratio $\frac{\frac{5}{4}}{\frac{9}{8}} = \frac{5}{4} \div \frac{9}{8} = \frac{5}{4} \times \frac{8}{9} = \frac{10}{9} = t = 182 \cdot 4$ cents

E and F=ratio $\frac{\frac{4}{3}}{\frac{5}{4}} = \frac{4}{3} \div \frac{5}{4} = \frac{4}{3} \times \frac{4}{5} = \frac{16}{15} = s = 111 \cdot 7$ cents

etc.

From which

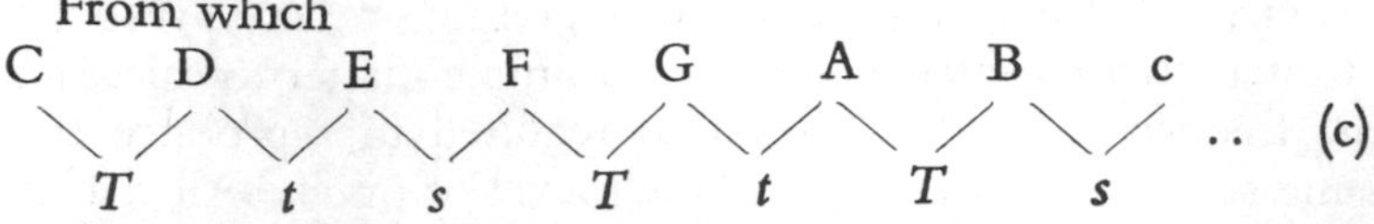

is obtained.

The set of ratios in group (b), which define the intervals of just temperament, are sometimes expressed as ratios without denominators, *i.e.* with 'understood' denominators of unity. This can be done by multiplying each ratio in group (b) by 24—the L.C.M. of the denominators—or, in this particular case, by dividing all the frequencies in group (a) by 10. This gives

C	D	E	F	G	A	B	c	
24	27	30	32	36	40	45	48	.. (a_1)

Now, reverting to the group of frequencies at (a)—suppose that D, not C, had been chosen as the reference note then the ratios so obtained would represent intervals made by the other notes with D. Proceeding on these lines by dividing group (a) through by the frequency corresponding with D—270—and expressing the ratios so obtained in their lowest terms yields the following set of ratios:—

C	D	E	F	G	A	B	c	
$\frac{8}{9}$	1	$\frac{10}{9}$	$\frac{32}{27}$	$\frac{4}{3}$	$\frac{40}{27}$	$\frac{5}{3}$	$\frac{16}{9}$	.. (e)

(This set of ratios can also be obtained directly from set (b) by multiplying it through by $\frac{8}{9}$.)

This result makes it clear that, from D's point of view, since neither the interval DF nor the interval DA are consonant, then either F and A, or D must be incorrectly placed—although, as already mentioned, these notes are in their correct positions for giving true concords on C, F and G. This explains why such a tuning, in fact any tuning, because it must temper some intervals, produces a temperament, not a scale. The reason for this 'discrepancy' can perhaps be best seen by examining the sequence of the diatonic steps between these notes as already available at (c). From this it can be seen that in terms of diatonic steps

the interval DF $= t + s$

whereas a true minor third $= T + s$ (*i.e.* EG or Ac)

also the interval DA $= T + 2t + s$
whereas a true fifth $= 2T + t + s$ (*i.e.* CG or Fc)

Comparing these equations it becomes apparent that if one of the t's in each of these intervals is replaced by a T then both intervals will be made true. Since T is greater than t by a comma ($T = t + k$) this amounts to widening each interval by a comma (ratio $\frac{81}{80}$). This may be done arithmetically as shown below:—

raising F by a comma gives ratio
$\frac{32}{27} \times \frac{81}{80} = \frac{6}{5}$ —corresponding to a true minor third

raising A by a comma gives ratio
$\frac{40}{27} \times \frac{81}{80} = \frac{3}{2}$ —corresponding to a true fifth

alternatively, lowering D by a comma

i.e., making D's ratio $= \dfrac{1}{\frac{81}{80}} = 1 \div \frac{81}{80} = 1 \times \frac{80}{81} = \frac{80}{81}$

(this is equivalent to lowering D to $\frac{10}{9}$ in the set of ratios at (b).)

then the interval DF = ratio

$$\frac{\frac{32}{27}}{\frac{80}{81}} = \frac{32}{27} \div \frac{80}{81} = \frac{32}{27} \times \frac{81}{80} = \frac{6}{5}$$ —corresponding to a true minor third

also the interval DA = ratio

$$\frac{\frac{40}{27}}{\frac{80}{81}} = \frac{40}{27} \div \frac{80}{81} = \frac{40}{27} \times \frac{81}{80} = \frac{3}{2}$$ —corresponding to a true fifth

Hence, in a major scale, even when no modulation is required, the second degree, or possibly the fourth and sixth degree notes—depending on the musical circumstances—must be mutable notes if the concords on the second degree are to sound in tune. In the same way, when considering the minor scale, at least the second and seventh degree notes are required to be mutable (see Figs. C and D), while, in modal music even more notes are required to be mutable. Thus a flexible intonation is demanded by the pure scale in order that all its essential intervals may sound in tune. It must not be confused with just temperament.

Pythagorean Tuning

In the previous work on just temperament it was found that one of its fifth degree intervals—DA—was a comma short of a pure fifth. The fifth being, after the octave, the most important interval in musical theory, it is now proposed to examine systematically, the intervals between all notes five degrees apart in just temperament, starting from F. This can best be done by first writing out the ratios corresponding to three or four consecutive octaves of just temperament—the ratios for each octave after the first being obtained by multiplying the corresponding ratios of the previous octave by 2—and then extracting every fifth ratio as follows:—

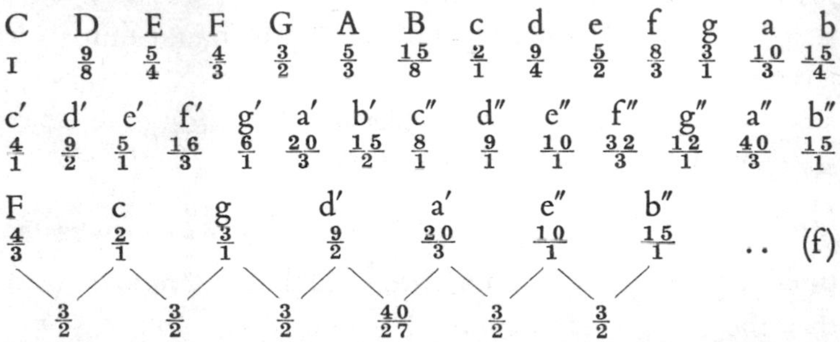

The first three of these intervals, Fc, cg and gd′ were made true fifths by choice in establishing just temperament, so, indirectly, were the last two, a′e″ and e″b″. For, in forming the major triad FAc, Ac was made a minor third, while, in making the major triad CEG, CE was made a major third. Hence any E must make, with the A below it, a perfect fifth. Again, since in the same process, EG was made a minor third and GB a major third, it follows that any B must make, with the E below it, a perfect fifth. The remaining interval d′a′, ratio $\frac{40}{27}$, has already been encountered in the set of ratios at (e)—where it was found to be a comma (ratio $\frac{81}{80}$) short of a perfect fifth (ratio $\frac{3}{2}$).

Now supposing it to be essential, above all other considerations, that these intervals all be made perfect fifths, then it would first be necessary to increase the interval d′a′ by a comma, by replacing a′ by a′ + k; this, in turn, would make it necessary to replace the note e″ by e″ + k, and b″, by b + k.

The ratios corresponding to these new notes are as follows:—

$$a' + k = \frac{20}{3} \times \frac{81}{80} = \frac{27}{4}$$
$$e'' + k = \frac{10}{1} \times \frac{81}{80} = \frac{81}{8}$$
$$b'' + k = \frac{15}{1} \times \frac{81}{80} = \frac{243}{16}$$

This gives, for the succession of perfect fifths, the following ratios:—

F	c	g	d′	a′+*k*	e″+*k*	b″+*k*	
$\frac{4}{3}$	$\frac{2}{1}$	$\frac{3}{1}$	$\frac{9}{2}$	$\frac{27}{4}$	$\frac{81}{8}$	$\frac{243}{16}$	(g)

In passing, since these notes form a succession of perfect fifths, it follows that their corresponding ratios can be expressed in terms of successive powers of $\frac{3}{2}$. This can best be shown from the above ratios by first multiplying each of them by $\frac{3}{4}$, as follows:

$1 \quad \frac{3}{2} \quad \frac{9}{4} \quad \frac{27}{8} \quad \frac{81}{16} \quad \frac{243}{32} \quad \frac{729}{64}$

i.e.

$(\frac{3}{2})^0 \quad (\frac{3}{2})^1 \quad (\frac{3}{2})^2 \quad (\frac{3}{2})^3 \quad (\frac{3}{2})^4 \quad (\frac{3}{2})^5 \quad (\frac{3}{2})^6$

When reduced, for ease of comparison, so as to lie within the ratios of 1 (or $\frac{1}{1}$) and $\frac{2}{1}$, the new ratios at (g) become

$$A+k = \frac{27}{4} \times \frac{1}{4} = \frac{27}{16}$$
$$E+k = \frac{81}{8} \times \frac{1}{8} = \frac{81}{64}$$
$$B+k = \frac{243}{16} \times \frac{1}{8} = \frac{243}{128}$$

Reducing to 'standard form', in similar manner, all the other notes in this succession of fifths, gives the following group of ratios—known as *Pythagorean Tuning*

C	D	E+*k*	F	G
1	$\frac{9}{8}$	$\frac{81}{64}$	$\frac{4}{3}$	$\frac{3}{2}$
0	203·9	407·8	498·0	702·0
1000	888·8	790·1	750·0	666·7
260·74	293·33	330·00	347·65	391·11

A+*k*	B+*k*	c	
$\frac{27}{16}$	$\frac{243}{128}$	$\frac{2}{1}$	.. .. (h)
905.8	1109·8	1200	cents
592·6	526·7	500·0	string-lengths in mm
440·00	495·00	521·48	frequencies based on a—440

A comparison of the steps between adjacent ratios in this tuning with those obtained previously at (c) for just temperament, shows that, as a result of the derivation of these ratios from those forming a succession of perfect 'fifths', four of the steps between them have had to be shifted by a comma—two, EF and Bc reduced, and two, DE and GA, increased from a minor tone to a major tone. The total change over an octave is zero—which is as it should be.

The ratio corresponding to the new interval may be calculated as shown below:—

$$s - k = \frac{\frac{16}{15}}{\frac{81}{80}} = \tfrac{16}{15} \div \tfrac{81}{80} = \tfrac{16}{15} \times \tfrac{80}{81} = \tfrac{256}{243} = 90{\cdot}225 \text{ cents}$$

The steps of Pythagorean tuning are thus arranged as follows:—

C		D		E+k		F		G		A+k		B+k		c.. (i)
	T		T		$s-k$		T		T		T		$s-k$	
	$\frac{9}{8}$		$\frac{9}{8}$		$\frac{256}{243}$		$\frac{9}{8}$		$\frac{9}{8}$		$\frac{9}{8}$		$\frac{256}{243}$	

Note that in this tuning there is only one size of whole tone—the major tone T, ratio $\frac{8}{9}$.

The ratio $s - k$ is known as the Pythagorean limma, or Greek hemitone (h).

Thus $h = s - k$ (u)

There is a strong tendency for singers, violinists, etc., to produce an approximation to this type of intonation in unaccompanied melody.

Regular Temperaments

Any regular temperament may, by definition, be derived from the ratios 'generated' by a succession of identical 'fifths' (*i.e.* ratios corresponding to fifths)—whether these be sharpened, flattened or true.

Taking initially the particular case where it is required to derive—say—a simple major diatonic tuning from the minimum number of such a succession of 'fifths', then, the 1st, 2nd, 3rd, 4th and 5th of these 'fifths' above the 'starting' or 'reference note' chosen—which will correspond to the tonic—must, when suitably transposed to the same octave, be made to correspond to the 5th, 2nd, 6th, 3rd and 7th degrees of the temperament. The 4th degree will correspond to that ratio which lies one 'fifth' below the chosen 'reference note'. Thus the series virtually start from one 'fifth' below the tonic. See Table A.

These facts can be stated in symbols as indicated below:—

$T = 2(5\text{th}) - 8\text{th}$
$\quad = 2(5\text{th}) - [5\text{th} + 4\text{th}]$
$\quad = 5\text{th} - 4\text{th}$ (hr)

$$M6 = 3(5\text{th}) - 8\text{th}$$
$$= 5\text{th} + [2(5\text{th}) - 8\text{th}]$$
$$= 5\text{th} + T \qquad \text{(or)}$$
$$M3 = 4(5\text{th}) - 2(8\text{th})$$
$$= 2[2(5\text{th}) - 8\text{th}]$$
$$= 2T \qquad \text{(kr)}$$
$$s = 3(8\text{th}) - 5(5\text{th})$$
$$= 3(8\text{th}) - 5\text{th} - 4(5\text{th})$$
$$= 3(8\text{th}) - 5\text{th} - [2(8\text{th}) + M3] \text{ from '}M3\text{' eq. above.}$$
$$= 8\text{th} - 5\text{th} - M3$$
$$= 4\text{th} - M3 \text{ from eq. (at), or eq. (ar) below} \qquad \text{(ir)}$$
$$4\text{th} = -5\text{th} + 8\text{th} \qquad \text{(at) or}$$

etc. (ar)

Given a true octave and the value of the ratio corresponding to any other interval in a regular temperament, then the above equations will enable the value of the ratios of all the remaining intervals to be calculated.

Alternatively, since in such a system there will be only one value corresponding to the whole tone, *T*, it follows that equations (j) to (p) can also be used to calculate the remaining ratios provided that 'T' and 't' are replaced by '*T*', and the other symbols by their corresponding equivalents.

The use of more than six successive '5ths' to obtain the various chromatic intervals, or to cater for modulation, will also, in general, make available certain 'odd' intervals—'odd' in the sense that, though differing from the 'standard' intervals established 'diatonically' by the first six '5ths', they may yet be sufficiently like these 'standard' intervals as to be heard as different versions of them.

For example, if the '5ths' are all true (Pythagorean Tuning) then the ratio between the extreme 'notes' of—say—any eight of these '5ths' (F—C♯, E♭—B, etc.) when referred to the same octave, will be ratio $\frac{6561}{4096}$, or 816 cents, and its inversion $\frac{8192}{6561}$ or 384 cents—practically a true minor sixth and a true major third. Whereas, all the 'standard' minor sixths and major thirds established previously among the first six '5ths' between all 'notes' four '5ths' apart, are of ratio, $\frac{128}{81}$, or 792 cents, and $\frac{81}{64}$, or 408 cents. Again, in the same temperament, the ratio between the extreme 'notes' of any nine of these '5ths' (F—G♯, A♭—B, etc.), when referred to the same octave, will be ratio $\frac{19683}{16384}$, or 318 cents, and its inversion, $\frac{32768}{19683}$, or 882 cents—practically a true minor third and a true major sixth. Whereas, all the previously established 'standard' major sixths and minor thirds which occur in this temperament between all 'notes' three '5ths' apart are of ratio, $\frac{27}{16}$, or 906 cents, and $\frac{32}{27}$, or 294 cents—and so on.

Basic Mean-tone Temperament

Since any four adjacent true fifths produce a true major third sharpened by a comma (plus two octaves), it follows that if the major thirds and not the fifths are to be kept true then the sum of every four adjacent fifths must be reduced by a comma. This objective is achieved in an abrupt and crude manner in just temperament by taking the whole of this comma from one and the same fifth, whereas, in *Basic Mean-Tone Temperament* a smoother and more subtle method is adopted—all fifths being reduced by a quarter of a comma thus making this system a regular one.

There are many ways of arriving at the ratios of an octave of this tuning. Probably the most obvious, in view of the angle from which the idea has been developed above, is to start with the succession of fifths—obtained already at (g)—systematically reduce each fifth by $k/4$, and finally telescope all the ratios obtained in this way so as to make them fall within a single octave.

Proceeding on these lines,

F	c	g	d′	a′+k	e″+k	b″+k
$\frac{4}{3}$	$\frac{2}{1}$	$\frac{3}{1}$	$\frac{9}{2}$	$\frac{27}{4}$	$\frac{81}{8}$	$\frac{243}{16}$

Reducing every fifth by $k/4$, starting, for convenience, from c

F+$k/4$	c	g−$k/4$	d′−$k/2$			
i.e. F+$k/4$	c	g−$k/4$	d′−$k/2$			
				a′+k−$3k/4$	e″+k−k	b″+k−$5k/4$
				a′+$k/4$	e″	b″−$k/4$ . . (j)

Finally, reducing these to one octave and arranging them in order gives

C		$D-k/2$		E		$F+k/4$		$G-k/4$		$A+k/4$		$B-k/4$		c	(k)
	$T-\frac{k}{2}$ or $t+\frac{k}{2}$		$t+\frac{k}{2}$ or $T-\frac{k}{2}$		$s+\frac{k}{4}$		$T-\frac{k}{2}$ or $t+\frac{k}{2}$		$t+\frac{k}{2}$ or $T-\frac{k}{2}$		$T-\frac{k}{2}$ or $t+\frac{k}{2}$		$s+\frac{k}{4}$		

The steps between adjacent ratios may be obtained by comparing them with the corresponding steps as obtained previously for just temperament at (c).

In just temperament	therefore	*In mean-tone tuning*
$CD=T$		$C(D-k/2)=T-k/2=203{\cdot}9-\frac{21{\cdot}5}{2}$ $=193{\cdot}15$ cents
$DE=t$		$(D-k/2)E=t+k/2=182{\cdot}4+\frac{21{\cdot}5}{2}$ $=193{\cdot}15$ cents
$EF=s$		$E(F+k/4)=s+k/4=111{\cdot}73+\frac{21{\cdot}5}{4}$ $=117{\cdot}1$ cents

etc.

Now, since $k=T-t$, therefore

$$T-\frac{k}{2}=T-\frac{T-t}{2}=\frac{2T-T+t}{2}=\frac{T+t}{2}$$

—in logarithmic units $=\dfrac{203{\cdot}9+182{\cdot}4}{2}=193{\cdot}15$ cents

also

$$t+\frac{k}{2}=t+\frac{T-t}{2}=\frac{2t+T-t}{2}=\frac{T+t}{2}$$

— in logarithmic units $=\dfrac{203{\cdot}9+182{\cdot}4}{2}=193{\cdot}15$ cents

So that this system of tuning requires only one size of whole tone whose ratio is the geometric mean of the major and minor tones—hence the name mean-tone.

This suggests another method by which mean-tone ratios may be derived. For if all the whole tones are to be equal and the major thirds true, then two such tones must equal a major third. Further, if the octave is to remain true, then two semitones of this system must equal an octave minus five of these whole tones.

Thus

2 mean-tones $= M3$, *i.e.* ratio $\frac{5}{4}$

1 mean-tone $= \frac{M3}{2}$, *i.e.* ratio $\sqrt{\frac{5}{4}} = \frac{\sqrt{5}}{2} = \frac{2\cdot23607}{2}$

$= 1\cdot118035$ say $1\cdot118 = 193$ cents

and

2 mean-tone semitones

$= 1$ octave $- 5$ mean-tones

$=$ ratio, $\frac{2}{1} \div (\frac{5}{4} \times \frac{5}{4} \times \frac{\sqrt{5}}{2})$ since $\frac{5}{4}$ represents 2 whole tones of mean-tone.

$=$ ratio $\frac{2}{1} \times \frac{4}{5} \times \frac{4}{5} \times \frac{2}{\sqrt{5}} = \frac{64}{25\sqrt{5}} = \frac{64}{25 \times 2\cdot23607}$

$\therefore$ the ratio of 1 mean-tone semitone

$$= \sqrt{\frac{64}{25 \times 2\cdot23607}} = \frac{8}{5 \times \sqrt{2\cdot23607}} = \frac{1\cdot6}{1\cdot49535}$$

$$= 1\cdot06998$$

say $1\cdot07 = 117\cdot1$ cents.

The remaining mean-tone ratios may then be obtained by a building-up process from these ratios—as follows,

mean-tone 4th $= 2$ mean-tones $+$ mean-tone semitone

$=$ ratio $1\cdot118 \times 1\cdot118 \times 1\cdot07 = 1\cdot3375$

or $= M3 +$ mean-tone semitone

$=$ ratio $\frac{5}{4} \times 1\cdot07 = 1\cdot25 \times 1\cdot07 = 1\cdot3375$

However, those who are familiar with the use of fractional powers, etc., may find it an advantage to express all mean-tone ratios in terms of powers or roots of 5, and then evaluate them from the list of roots or fractional powers of 5 given on page 300.

For example,

the ratio of a mean-tone fifth $= \sqrt{\sqrt{5}} = 5^{\frac{1}{4}} = 1\cdot49535$

the ratio of a mean-tone sixth $= \sqrt[4]{\frac{5^3}{2}} = \frac{5^{\frac{3}{4}}}{2} = \frac{3\cdot34371}{2}$

$= 1\cdot67185$

etc.

Finally, starting with the knowledge that every fifth is to

be reduced by an amount $k/4$ and that every major third is to be kept true, the ratios of this tuning can be obtained by considering major triads on the tonic, dominant and subdominant—as was done in the case of just temperament.

Thus the particulars of Basic Mean-Tone Temperament are as follows:

C	D—$k/2$	E	F+$k/4$	G—$k/4$
1·0000	1·1180	1·2500	1·3375	1·4953
0	193·16	386·31	503·42	696·58
1000·0	894·5	800·0	747·7	668·7
263·18	294·25	328·98	352·00	393·55

A+$k/4$	B—$k/4$	c	.. .. (I)
1·6719	1·8692	2·0000	ratios
889·74	1082·9	1200·0	cents
598·1	535·0	500·0	string-lengths in mm
440·00	491·93	526·36	frequencies based on a — 440

Provision of sets of diatonic intervals at more than one pitch
Cyclic temperaments   *Cycles of* 53, 50, 31, 19, *and* 12
Equal temperament

If it is required to repeat, at some other pitch, a set of diatonic intervals to one or other of the tunings that have been previously dealt with, then the frequency ratios of the additional 'notes' required may be derived from an extension, in either or both directions, of the series of fifths peculiar to that tuning or temperament—the assumptions being the same as for the original set. This has been done in Table A for just, Pythagorean and mean-tone tunings for a limited number of five extra 'notes' at each end of the appropriate series—although the normal keyboard can only accommodate a total of twelve such notes at any one time.

In Figs. C and D are shown a 'folded back' chain of pure fifths extending to F♯, in the one direction, and to B♭ in the other.

If the additional notes are found to be suitable for chromatic usage—then so much the better.

Since the numbers 3 and 2 (and 5) are prime to one another, it follows that no exact number of true fifths, ratio $\frac{3}{2}$, (or major thirds, ratio $\frac{5}{4}$) can ever fit exactly into an exact number of octaves, ratio $\frac{2}{1}$, so that, at least in theory, the series of true fifths—such as is used to derive Pythagorean tuning—may be extended indefinitely, as also its notation, in either direction without a single unison occurring with successive octaves tuned from the same note. However, a few 'near misses' do occur, where—if the discrepancy when shared out equally among all the fifths concerned is sufficiently small as to leave more or less unimpaired the ear's estimation of them as true intervals—'fifths' and octaves are effectively fitted and the temperament or tuning becomes cyclic, *i.e.* the octave is divided into a number of equal parts enabling endless modulation through a cycle of fifths to be made from a limited number of notes. The latter demands an enharmonic change somewhere in the circle, both in notation—so as not to run out of it—and also presumably to some extent in pure aural perception, as no change in frequency of any of the 'notes' occurs. In practice this kind of tuning makes fifths, either above or below any of its notes, available with equal facility.

The problem of locating these 'near misses' can be solved simply by listing the number of cents (or other logarithmic units) in 1, 2, 3, etc. fifths, and then comparing these with the number of cents in 1, 2, 3, etc. octaves. However, this method becomes tedious if a search is required to be made for fairly accurate ratios over a wide range of values and in this case another method is recommended in which the number of fifths in one octave is expressed, as near as is possible, as a ratio of two reasonably low whole numbers (*e.g.* $\frac{12}{7}$ in equal temperament). These numbers then give, to a degree of approximation which can be determined, a whole number of 'fifths' that equal, more or less, a whole number of octaves. Some ratios of fifths to octaves obtained in this manner are given below in order of accuracy, together with their corresponding deviations from a true fifth quoted in brackets, in cents—assuming a true octave. For details of the method see Appendix 9.

The ratio of the intervals $\dfrac{\text{Octave (in cents)}}{\text{Fifth (in cents)}}$

$= \frac{1200}{701 \cdot 955} \simeq \quad \frac{19}{11} \qquad \frac{50}{29} \qquad \frac{31}{18}$

$$\begin{array}{cccl} (-7{\cdot}219) & (-5{\cdot}955) & (-5{\cdot}181) & \\ \frac{12}{7} & \frac{41}{24} & \frac{53}{31} & \text{etc.} \quad .. \quad \text{(m)} \\ (-1{\cdot}955) & (+{\cdot}484) & (-{\cdot}068) & \end{array}$$

(The sign '≏' means 'is approximately equal to'.)

But thirds (and sixths) must also be considered. A division which gives a reasonably good fifth may give an unsuitable third—and vice-versa. The application of the same method as above yields the ratios $\frac{3}{1}$, $\frac{28}{9}$ and $\frac{59}{19}$ as the first three successive approximations to the ratio octave/major third shown below. The ratio $\frac{3}{1}$ may also be expressed as $\frac{6}{2}$ or $\frac{9}{3}$ or $\frac{12}{4}$, etc., and additional ratios may be obtained by combining ratios as demonstrated in Appendix 9, bearing in mind that the only ratios worth considering, in this particular case, are those which have, or can be made to have, the same numerators ('tops') as those already obtained at (m) above.

The ratio of the intervals $\frac{\text{Octave (in cents)}}{\text{Major Third (in cents)}}$

$$= \frac{1200}{386{\cdot}314} \fallingdotseq \quad \frac{12}{4} \quad \frac{19}{6} \quad \frac{50}{16}$$
$$(+13{\cdot}69) \quad (-7{\cdot}367) \quad (-2{\cdot}314)$$

$$\begin{array}{ccl} \frac{53}{17} & \frac{31}{10} & \text{etc.} \ldots \quad .. \quad .. \quad .. \quad .. \quad \text{(n)} \\ (-1{\cdot}408) & (+{\cdot}783) & \end{array}$$

These two sets of ratios confirm the basis of at least five recognised cyclic temperaments.

(Suitability for melody, a most important factor, has been entirely ignored here. This is even more difficult to satisfy (ideally) than is pure (harmonic) intonation—from a limited number of notes of fixed intonation.)

The ratios $\frac{53}{31}$ and $\frac{53}{17}$ suggest splitting the octave into 53 equal parts or degrees and taking 31 of them for the fifth and 17 of them for the major third. This is known as the cycle of 53 and gives practically pure fifths and thirds. It was for this tuning that Bosanquet, during the years 1872-3, designed and made a generalised fingerboard which is still in the possession of the Science Museum at South Kensington.

As already mentioned (Chapter 15), Dr. Robert Smith devised a system of temperament ('System of Equal Harmony') based not on mathematical jugglery with the exact ratios corresponding to the consonant intervals of the fifth, major third, etc., but on his own estimation of the demands of the musical ear. He perceived that it was in moving from better harmony to worse that the ear was most offended, so that it was useless to make one (or more) consonances true (or very nearly so) at the expense of the others. What was required, he concluded, was to make all consonances as equally harmonious as possible. He therefore proceeded to develop a temperament in which the dissonance brought about by any tempering of the consonant intervals was distributed equally between the important concords. In order to accomplish this he assumed that the tempered intervals would be equally harmonious if they produced beats (*i.e.* between those pairs of upper partials whose unison defines the interval) at the same rate in adjacent consonant intervals (such as the fifth and major sixth), and accordingly, in his system, made the fifth, major sixth and major third, deviate from their true values by $-\frac{5}{18}$, $+\frac{3}{18}$, and $-\frac{2}{18}$ of a comma, respectively (*i.e.* −5·97, +3·58, and −2·31 cents respectively). He also showed that this system was almost identical with that which divided the octave into 50 equal parts ($\frac{1200}{50}=24$ cents each), taking 29 of these for the fifth ($29\times24=696$ cents) and 16 for the major third ($16\times24=384$ cents), see Fig. H. Under these circumstances the deviations mentioned above work out to −5·96, + 3·64, and −2·31 cents respectively. The intervals between the notes of this cyclic temperament may be obtained by expressing the temperament as a succession of its fifths (696 cents each)—as has been done for some other temperaments in Table A—near the end of this chapter.

The ratios $\frac{50}{29}$ and $\frac{50}{16}$, corresponding to the above division, occur in the sets of ratios at (*m*) and (*n*).

The ratios $\frac{31}{18}$ and $\frac{31}{10}$ suggest the splitting of the octave into 31 degrees—the fifth into 18 degrees, and the major third into 10 degrees. This gives the (Huygens) cycle of 31, having a fifth about a quarter of a comma (−5.181 cents) flat and thus almost identical with the mean-tone fifth (−5.377 cents), and a practically pure major third (+ .783 cents). The 31 notes of this cyclic system thus provide what is virtually an extended form of meantone in which all keys are equally available without 'wolves'. Much the same also applies to the previous 50 cycle system.

The ratios $\frac{19}{11}$ and $\frac{19}{6}$ suggest the splitting of the octave into 19 degrees, taking 11 of them for the fifth, and 6 of them for the major third. This is the cycle of 19—sometimes associated with Woolhouse. Its fifths are rougher than any of the other systems considered here.

Finally, the ratios $\frac{12}{7}$ and $\frac{12}{4}$ suggest the splitting of the octave into 12 degrees and allocating 7 of these for the fifth and 4 of them for the major third. This gives a cycle of 12 usually called equal temperament in which the fifth is relatively good but the major third is decidedly bad, and the major sixth worse still—at least from the point of view of consonance—which is the only criterion being considered here.

In order to make some comparison of these five cyclic temperaments it is instructive to make out a list of intervals and the number of degrees allocated to them in the respective temperaments as shown below:—

octave (8th)		53	50	31	19	12
fifth (5th)		31	29	18	11	7
major third (*M*3)		17	16	10	6	4
fourth (4th)	8th–5th	22	21	13	8	5
minor third (*m*3)	5th–*M*3	14	13	8	5	3
major sixth (*M*6)	8th–*m*3	39	37	23	14	9
minor sixth (*m*6)	8th–*M*3	36	34	21	13	8
major tone (*T*)	5th–4th	9	8	5	3	2
minor tone (*t*)	*M*6–5th	8	8	5	3	2
diatonic semitone (*s*)	4th–*M*3	5	5	3	2	1
larger chromatic semitone	*T*–*s*	4	3	2	1	1
smaller chromatic semitone	*t*–*s*	3	2	2	1	1
comma (*k*)	*T*–*t*	1	0	0	0	0

Here it can be seen that, among other things, the cycle of 53 is the only cyclic temperament in which a distinction between the larger and the smaller chromatic semitone, and between the major and minor tones, is possible—and then only if used in conjunction with a most elaborate keyboard. With the cycles of 50, 31 and 19 it is possible to distinguish the diatonic semitone from a chromatic semitone (of a sort), but again, only by using an elaborate keyboard—unless one is content to restrict the immediately available range by employing a standard keyboard in conjunction with a set of special stops or switches (see equal temperament, etc., Chapter 15). Finally, equal temperament makes all its semitones the same and exactly equal to half of one of its whole tones and provides an unrestricted range through those keys which have now become associated with the standard keyboard—but at the price of harmonically poor, though tolerably melodic, thirds and sixths.

In a cyclic temperament of n degrees, the value of one degree may be obtained by splitting the octave into n equal parts, so that,

in cents,

$$1 \text{ degree} = \frac{1200}{n} \qquad 2 \text{ degrees} = \frac{1200}{n} \times 2, \text{ and so on,}$$

as a ratio

$$1 \text{ degree} = \left(\frac{2}{1}\right)^{\frac{1}{n}} = \frac{(2)^{\frac{1}{n}}}{1} = \frac{\sqrt[n]{2}}{1}$$

$$2 \text{ degrees} = \left(\frac{2}{1}\right)^{\frac{2}{n}} = \frac{(2)^{\frac{2}{n}}}{1} = \frac{(\sqrt[n]{2})^{2}}{1}, \text{ and so on.}$$

These two different measures of the same interval, are, of course, related as follows.

Denoting the ratio of the interval by r and the logarithmic measure of the (physical) interval by—say—I, then,

$$r = \text{antilog}_{10} \text{ of } \frac{\text{number of cents in the interval}}{4000} = \text{antilog}_{10} \frac{I}{4000}$$

Conversely,
the number of logarithmic units corresponding to ratio r, is given by $I = 4000 \log_{10} r$ cents.

If greater accuracy than this is required 3986 should be substituted for 4000 in the above formulae.

Thus in the cycle of 53,

$$1 \text{ degree, in cents} = \frac{1200}{53} = 22{\cdot}64 \; (22{\cdot}64151)$$

$$\text{as a ratio} = \frac{\sqrt[53]{2}}{1} = \frac{2^{\frac{1}{53}}}{1} = \frac{1\cdot013}{1}\left(\frac{1\cdot013164}{1}\right)$$

$$= (\text{approx.})\ \frac{77}{76}$$

∴ 31 degrees, or

$$1 \text{ fifth, in cents} = 31 \times 22\cdot64 = 701\cdot9\ (701\cdot887)$$

$$\text{as a ratio} = \frac{(\sqrt[53]{2})^{31}}{1} = \frac{2^{\frac{31}{53}}}{1} = \frac{1\cdot5}{1}\left(\frac{1\cdot49994}{1}\right)$$

$$= (\text{approx.})\ \frac{12499}{8333}$$

In the cycle of 50,

$$1 \text{ degree, in cents} = \frac{1200}{50} = 24$$

$$\text{as a ratio} = \frac{\sqrt[50]{2}}{1} = \frac{2^{\frac{1}{50}}}{1} = \frac{1\cdot014}{1}\left(\frac{1\cdot01396}{1}\right)$$

$$= (\text{approx.})\ \frac{145}{143}$$

$$\therefore 29 \text{ degrees, in cents} = 29 \times 24 = 696$$

$$\text{as a ratio} = \frac{(\sqrt[50]{2})^{29}}{1} = \frac{2^{\frac{29}{50}}}{1} = \frac{1\cdot495}{1}\left(\frac{1\cdot49485}{1}\right)$$

$$= (\text{approx.})\ \frac{3338}{2233}$$

In the cycle of 31,

$$1 \text{ degree, in cents} = \frac{1200}{31} = 38\cdot71\ (38\cdot709677)$$

$$\text{as a ratio} = \frac{\sqrt[31]{2}}{1} = \frac{(2)^{\frac{1}{31}}}{1} = \frac{1\cdot023}{1}\left(\frac{1\cdot022611}{1}\right)$$

$$= (\text{approx.})\ \frac{181}{177}$$

∴ 18 degrees, or

$$1 \text{ fifth, in cents} = 18 \times 38\cdot71 = 696\cdot8\ (696\cdot774)$$

$$\text{as a ratio} = \frac{(\sqrt[31]{2})^{18}}{1} = \frac{2^{\frac{18}{31}}}{1} = \frac{1\cdot496}{1}\left(\frac{1\cdot49552}{1}\right)$$

$$= (\text{approx.})\ \frac{166}{111}$$

In the cycle of 19,

$$1 \text{ degree, in cents} = \frac{1200}{19} = 63{\cdot}16\ (63{\cdot}157895)$$

$$\text{as a ratio} = \frac{\sqrt[19]{2}}{1} = \frac{(2)^{\frac{1}{19}}}{1} = \frac{1{\cdot}037}{1}\ \left(\frac{1{\cdot}03716}{1}\right)$$

$$= (\text{approx.})\ \frac{307}{296}$$

∴ 11 degrees, or

$$1 \text{ fifth, in cents} = 11 \times 63{\cdot}16 = 694{\cdot}7\ (694{\cdot}737)$$

$$\text{as a ratio} = \frac{(\sqrt[19]{2})^{11}}{1} = \frac{2^{\frac{11}{19}}}{1} = \frac{1{\cdot}494}{1}\ \left(\frac{1{\cdot}49376}{1}\right)$$

$$= (\text{approx.})\ \frac{118}{79}$$

In the cycle of 12,

$$1 \text{ degree, in cents} = \frac{1200}{12} = 100$$

$$\text{as a ratio} = \frac{\sqrt[12]{2}}{1} = \frac{2^{\frac{1}{12}}}{1} = \frac{1{\cdot}059}{1}\ \left(\frac{1{\cdot}059463}{1}\right)$$

$$= (\text{approx.})\ \frac{89}{84}$$

∴ 7 degrees, or

$$1 \text{ fifth, in cents} = \frac{1200}{12} \times 7 = 700$$

$$\text{as a ratio} = \frac{(\sqrt[12]{2})^{7}}{1} = \frac{2^{\frac{7}{12}}}{1} = \frac{1{\cdot}498}{1}\ \left(\frac{1{\cdot}498307}{1}\right)$$

$$= (\text{approx.})\ \frac{442}{295}$$

The more exact figures given in brackets above are quoted purely as a matter of interest, although it should be noted that, if in estimating the intervals of—say—the cycle of 53 by first finding the value of one degree, then, an error of such a small amount as ·1 of a cent in this value would result in an error of 3·1 cents (31 × ·1) in each fifth, and a maximum error of 53 times this amount, *i.e.*, 164·3 cents (more than one and a half semitones of equal temperament) if a succession of 53 fifths was used.

Figure H opposite gives a visual indication of the relative sizes of the intervals used in these temperaments and tunings. In this diagram E denotes equal temperament, J—just temperament, P—Pythagorean tuning, M—mean-tone temperament, and 19, 31, 50 and 53—the cyclic temperaments employing these numbers of equal divisions to the octave. With the exception of just temperament, all these tunings are regular, *i.e.* each uses the same size fifth throughout. Equal temperament is also cyclic, employing a cycle of 12. The position of

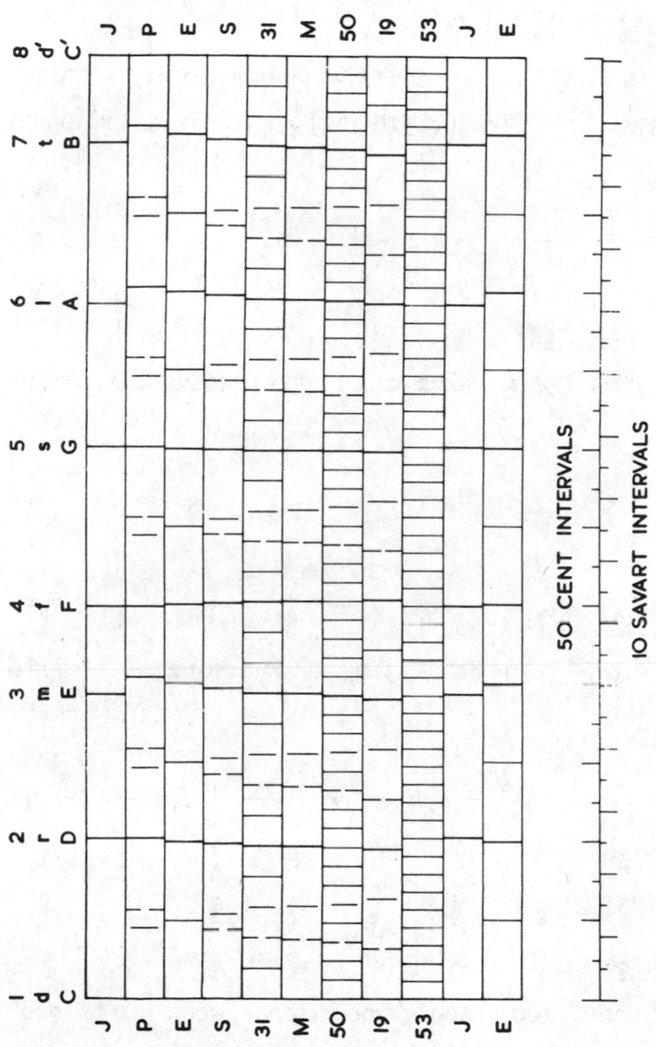

TABLE A. TEMPERAMENTS OR TUNINGS

JUST

	$G\flat_{III}$	$D\flat_{II}$	$A\flat_{II}$	$E\flat_{I}$ ★	$B\flat_{I}$	F	c	g	d′ ★
←etc.	$\frac{4}{45}$	$\frac{2}{15}$	$\frac{1}{5}$	$\frac{3}{10}$	$\frac{4}{9}$	$\frac{2}{3}$	$\frac{1}{1}$	$\frac{3}{2}$	$\frac{9}{4}$
	$\frac{64}{45}$	$\frac{16}{15}$	$\frac{8}{5}$	$\frac{6}{5}$	$\frac{16}{9}$	$\frac{4}{3}$	$\frac{1}{1}$	$\frac{3}{2}$	$\frac{9}{8}$
	609·78	111·73	813·69	315·64	996·09	498·04	0	701·96	203·91
	375·46	281·60	422·40	316·80	469·3	352·00	264·00	396·00	297·00

PYTHAGOREA

←etc.	$-k$	$-k$	$-k$	$-k$	0	0	0	0	0
	$\frac{64}{729}$	$\frac{32}{243}$	$\frac{16}{81}$	$\frac{8}{27}$	$\frac{4}{9}$	$\frac{2}{3}$	$\frac{1}{1}$	$\frac{3}{2}$	$\frac{9}{4}$
	$\frac{1024}{729}$	$\frac{256}{243}$	$\frac{128}{81}$	$\frac{32}{27}$	$\frac{16}{9}$	$\frac{4}{3}$	$\frac{1}{1}$	$\frac{3}{2}$	$\frac{9}{8}$
	588·3	90·2	792·2	294·1	996·1	498·0	0	702·0	203·9

MEAN-TON

←etc.	$+\frac{k}{2}$	$+\frac{k}{4}$	0	$-\frac{k}{4}$	$+\frac{k}{2}$	$+\frac{k}{4}$	0	$-\frac{k}{4}$	$-\frac{k}{2}$
	$\frac{1}{5^{\frac{3}{2}}}$	$\frac{1}{5^{\frac{5}{4}}}$	$\frac{1}{5}$	$\frac{1}{5^{\frac{3}{4}}}$	$\frac{1}{5^{\frac{1}{2}}}$	$\frac{1}{5^{\frac{1}{4}}}$	$\frac{1}{1}$	$\frac{5^{\frac{1}{4}}}{1}$	$\frac{5^{\frac{1}{2}}}{1}$
	$\frac{16}{5^{\frac{3}{2}}}$	$\frac{8}{5^{\frac{5}{4}}}$	$\frac{8}{5}$	$\frac{4}{5^{\frac{3}{4}}}$	$\frac{4}{5^{\frac{1}{2}}}$	$\frac{2}{5^{\frac{1}{4}}}$	$\frac{1}{1}$	$\frac{5^{\frac{1}{4}}}{1}$	$\frac{5^{\frac{1}{2}}}{2}$
	620·5	117·1	813·7	310·3	1006·8	503·4	0	696·6	193·2

EQUAL

$-\frac{5}{11}k$	$-\frac{6}{11}k$	$-\frac{7}{11}k$	$-\frac{8}{11}k$	$+\frac{2}{11}k$	$+\frac{1}{11}k$	0	$-\frac{1}{11}k$	$-\frac{2}{11}k$
$\frac{1}{2^{\frac{7}{2}}}$	$\frac{1}{2^{\frac{35}{12}}}$	$\frac{1}{2^{\frac{7}{3}}}$	$\frac{1}{2^{\frac{7}{4}}}$	$\frac{1}{2^{\frac{7}{6}}}$	$\frac{1}{2^{\frac{7}{12}}}$	$\frac{1}{1}$	$\frac{2^{\frac{7}{12}}}{1}$	$\frac{2^{\frac{7}{6}}}{1}$
$\frac{16}{2^{\frac{7}{2}}}$	$\frac{8}{2^{\frac{35}{12}}}$	$\frac{8}{2^{\frac{7}{3}}}$	$\frac{4}{2^{\frac{7}{4}}}$	$\frac{4}{2^{\frac{7}{6}}}$	$\frac{2}{2^{\frac{7}{12}}}$	$\frac{1}{1}$	$\frac{2^{\frac{7}{12}}}{1}$	$\frac{2^{\frac{7}{6}}}{2}$
$\frac{2^{\frac{1}{2}}}{1}$	$\frac{2^{\frac{1}{12}}}{1}$	$\frac{2^{\frac{2}{3}}}{1}$	$\frac{2^{\frac{1}{4}}}{1}$	$\frac{2^{\frac{5}{6}}}{1}$	$\frac{2^{\frac{5}{12}}}{1}$	$\frac{1}{1}$	$\frac{2^{\frac{7}{12}}}{1}$	$\frac{2^{\frac{1}{6}}}{1}$
600	100	800	300	1000	500	0	700	200

...XPRESSED AS A SUCCESSION OF FIFTHS

...TEMPERAMENT

′	e″	b″	f♯‴	c♯⁗ *	g♯⁗	d♯′′′′′	a♯′′′′′	→ etc.	J-I
$\frac{..0}{..3}$	$\frac{5}{1}$	$\frac{15}{2}$	$\frac{45}{4}$	$\frac{135}{8}$	$\frac{25}{1}$	$\frac{75}{2}$	$\frac{225}{4}$		J-II
$\frac{..5}{..3}$	$\frac{5}{4}$	$\frac{15}{8}$	$\frac{45}{32}$	$\frac{135}{128}$	$\frac{25}{16}$	$\frac{75}{64}$	$\frac{225}{128}$		J-III
84·36	386·31	1088·27	590·22	92·18	772·63	274·58	976·54		J-IV
40·00	330·00	495·00	371·25	278·4375	412·50	309·375	464·0625		JV

...TUNING

$+k$	$+k$	$+k$	$+k$	$+k$	$+2k$	$+2k$	$+2k$	→ etc.	P-I
$\frac{..7}{..8}$	$\frac{81}{16}$	$\frac{243}{32}$	$\frac{729}{64}$	$\frac{2187}{128}$	$\frac{6561}{256}$	$\frac{19683}{512}$	$\frac{59049}{1024}$		P-II
$\frac{..7}{..6}$	$\frac{81}{64}$	$\frac{243}{128}$	$\frac{729}{512}$	$\frac{2187}{2048}$	$\frac{6561}{4096}$	$\frac{19683}{16384}$	$\frac{59049}{32768}$		P-III
..05·9	407·8	1109·8	611·7	113·7	815·6	317·6	1019·6		P-IV

TEMPERAMENT

$\frac{k}{4}$	0	$-\frac{k}{4}$	$-\frac{k}{2}$	$-\frac{3}{4}k$	0	$-\frac{k}{4}$	$-\frac{k}{2}$	→ etc.	M-I
$..^{\frac{3}{4}}$	$\frac{5}{1}$	$\frac{5}{1}^{\frac{5}{4}}$	$\frac{5}{1}^{\frac{3}{2}}$	$\frac{5}{1}^{\frac{7}{4}}$	$\frac{5}{1}^{2}$	$\frac{5}{1}^{\frac{9}{4}}$	$\frac{5}{1}^{\frac{5}{2}}$		M-II
$..^{\frac{3}{4}}$	$\frac{5}{4}$	$\frac{5}{4}^{\frac{5}{4}}$	$\frac{5}{8}^{\frac{3}{2}}$	$\frac{5}{16}^{\frac{7}{4}}$	$\frac{5}{16}^{2}$	$\frac{5}{32}^{\frac{9}{4}}$	$\frac{5}{32}^{\frac{5}{2}}$		M-III
89·7	386·3	1082·9	579·5	76·0	772·6	269·2	965·8		M-IV

...EMPERAMENT

$-\frac{8}{11}k$	$+\frac{7}{11}k$	$+\frac{6}{11}k$	$+\frac{5}{11}k$	$+\frac{4}{11}k$	$+\frac{14}{11}k$	$+\frac{13}{11}k$	$+\frac{12}{11}k$	E-I
$\frac{2}{1}^{\frac{7}{4}}$	$\frac{2}{1}^{\frac{7}{3}}$	$\frac{2}{1}^{\frac{35}{12}}$	$\frac{2}{1}^{\frac{7}{2}}$	$\frac{2}{1}^{\frac{49}{12}}$	$\frac{2}{1}^{\frac{14}{3}}$	$\frac{2}{1}^{\frac{21}{4}}$	$\frac{2}{1}^{\frac{35}{6}}$	E-II
$\frac{2}{2}^{\frac{7}{4}}$	$\frac{2}{4}^{\frac{7}{3}}$	$\frac{2}{4}^{\frac{35}{12}}$	$\frac{2}{8}^{\frac{7}{2}}$	$\frac{2}{16}^{\frac{49}{12}}$	$\frac{2}{16}^{\frac{14}{3}}$	$\frac{2}{32}^{\frac{21}{4}}$	$\frac{2}{32}^{\frac{35}{6}}$	E-III
$\frac{2}{1}^{\frac{3}{4}}$	$\frac{2}{1}^{\frac{1}{3}}$	$\frac{2}{1}^{\frac{11}{12}}$	$\frac{2}{1}^{\frac{1}{2}}$	$\frac{2}{1}^{\frac{1}{12}}$	$\frac{2}{1}^{\frac{2}{3}}$	$\frac{2}{1}^{\frac{1}{4}}$	$\frac{2}{1}^{\frac{5}{6}}$	E-IIIa
900	400	1100	600	100	800	300	1000	E-IV

each sharpened note is indicated by a 'rung' broken at the centre, that of each flattened note by a 'rung' broken at each end.

Temperaments Expressed as a Succession of Fifths

In Table A the alphabetic notation in row J-I is used to represent the exact ratios of just temperament given in row J-II.

Rows P-I, M-I, S-I and E-I give the deviations of the listed temperaments from those of just temperament—in terms of the comma (k = ratio $\frac{81}{80}$, or 21·506 cents)—with any C as the reference note. To change this reference note see later in this chapter.

Each of the systems shown, except the first, is regular, *i.e.*, has identical fifths.

In just temperament, every fifth fifth is flattened by a comma. The remaining fifths are left true.

In the Pythagorean tuning all the fifths are true—ratio $\frac{3}{2}$ or 702 cents.

In equal temperament all the fifths are slightly flattened, their ratio being $\frac{2^{7/12}}{1}/1$ = 1·4983/1 or 700 cents.

In basic mean-tone temperament all the fifths are flattened by a quarter of a comma and thus have a ratio of $5^{1/4}/1$ = 1·4953/1 or 696·6 cents.

In just temperament every fifth fifth is flattened by a comma and so has a ratio of $\frac{40}{27}$. These flattened fifths, which are marked with an asterisk, make all the major thirds of this temperament true. The remaining fifths are true.

In the third row of each set are given the ratios 'generated' by the succession of fifths of the previous columns after their reduction to 'standard form'. The latter process consists of multiplying or dividing those ratios which lie outside the range $\frac{1}{1}$ to $\frac{2}{1}$ by that power of 2 which will bring them within this range.

The fourth row of each set expresses the ratios of the third row of that set in terms of cents above c. This figure taken from 1200 will give their corresponding inversions.

All the rows of just temperament except J-IV can be expressed *exactly* in figures and are so expressed in this table. However, row J-IV is given to more than usual accuracy in an attempt to pin-point its values and match—in logarithmic ratio units—the exactness of its other rows.

The row J-V gives the *exact* frequencies corresponding to the intervals of just temperament with a′ at 440 cycles. One way of calculating the frequencies to produce the intervals of the other temperaments is to use these figures in conjunction with the appropriate deviations and the Ratio Table as explained at the end of this chapter.

The Calculation of the Frequencies of a Temperament Using the Ratio Table

Once the set of frequency ratios corresponding to the intervals of a temperament is known and the frequency of its reference 'note' is chosen it becomes a relatively simple task to work out the frequencies corresponding to its remaining notes—as has already been explained earlier in this chapter—without having to refer to the Ratio Table.

If the ratios are known only in terms of cents then they may be converted to decimal ratios as indicated in Chapter 19, and hence to frequencies as above.

If the temperament is a regular one—with no 'odd', or 'alternative' intervals available—then only one of its particular intervals need be known and the remainder can be found. See later under 'Temperaments in General'. Alternatively, once the logarithmic measure of its tempered fifth is known it is a simple though rather tedious matter to express, in terms of this measured 'fifth', as many intervals of the temperament as are required, and, after reducing them to values lying between 0 and 1200 cents, to look them up in the Ratio Table and hence obtain their equivalent decimal ratios. This process is used in the calculations of the temperaments listed below. The values, in cents, of the 'downward fifths' are obtained by taking the values of the 'upward fifths' away from 1200. An 'upward fourth' is equivalent to a 'downward fifth'.

Cyclic temperaments that are regular can obviously be dealt with in the same way as regular temperaments—see, for example, Robert Smith's temperament treated as a cycle of

50 degrees with 29 degrees allocated to its 'fifth', as shown below.

	Silbermann's			Robert Smith's			
5ths	Cents	Ratio	Freq.	Cents	Ratio	Freq.	Note
0	0	1·00000	440·00	0	1·00000	440·00	a′
1	698·37	1·49690	329·32	696	1·49485	328·87	e′
2	196·74	1·12035	492·95	192	1·11729	491·61	b′
3	895·11	1·67705	386·95	888	1·67018	367·44	f′♯
4	393·48	1·25519	276·14	384	1·24833	247·63	c′♯
5	1091·85	1·87889	413·36	1080	1·86607	410·54	g′♯
6	590·22	1·40625	309·38	576	1·39474	306·84	d′♯
7	88·59	1·05251	463·10	72	1·04247	458·69	a′♯
8	786·96	1·57549	346·61	768	1·55833	342·83	e′♯
9	285·34	1·17918	518·84	264	1·16473	512·48	b′♯
etc.							etc.
−1	501·63	1·33610	293·94	504	1·33793	294·34	d′
−2	1003·26	1·78515	392·73	1008	1·79005	393·81	g′
−3	304·89	1·19257	262·37	312	1·19748	263·45	c′
−4	806·52	1·59339	350·55	816	1·60214	352·47	f′
−5	108·15	1·06446	468·36	120	1·07177	471·58	b′♭
−6	609·78	1·42222	312·89	624	1·43396	315·47	e′♭
−7	1111·41	1·90023	418·05	1128	1·91853	422·08	a′♭
−8	413·03	1·26944	279·28	432	1·28343	282·35	d′♭
−9	914·66	1·69610	373·14	936	1·71713	377·77	g′♭
etc.							etc.

Temperaments in General

The following equations must hold in *any* temperament in which the octaves are kept true,

$$4\text{th} + 5\text{th} = 8\text{th} \qquad \text{(at)}$$
$$m3 + M6 = 8\text{th} \qquad \text{(bt)}$$
$$M3 + m6 = 8\text{th} \qquad \text{(ct)} \qquad \text{etc.}$$

all symbols for intervals except '8th' being here used to represent their values in any particular temperament.

Such pairs of intervals are said to be *complementary* and corresponding equations will hold for all such pairs of complementary intervals.

If more than one 'version' of a particular interval occurs in a tuning then there will be an appropriate equation for each of them.

If, in addition to the above, the temperament is a regular one, then, first of all equations (hr) etc., under Regular Temperaments in this chapter, will apply for the conditions stated. Secondly, these can be used to show that equations (d) to (p) of Chapter 19 will also apply, in this case, to their corresponding tempered values. Finally, it will be seen that there can now be only one size of whole tone—which considerably simplifies equations (j) to (p)—and that it is therefore only possible to choose the value of one interval—other than the octave—the remaining intervals then being automatically fixed by these two values. As an example of this consider the particular case of equal temperament. Since it is regular we can deduce from eq. (k) that its major third will be such that,

$$M3 = 2T \qquad \text{(i)}$$

where T is an equally tempered tone. But the cyclic nature of equal temperament demands that there be exactly three of these major thirds to the octave (see Eq. (2), p. 53) hence,

$$\text{8th} = 3(M3) = 3(2T) = 6T \qquad \text{(ii)}$$

which fixes one equally tempered tone as exactly one-sixth of an octave—a fact which automatically establishes the widths of all the remaining intervals of the temperament. The equally tempered semitone will, for instance, have now to equal exactly half an equally tempered tone. For, eq. (p) can now be rewritten as,

$$\text{8th} = 5T + 2s \qquad \text{(iii)}$$

but eqs. (ii) and (iii) must be equal so we can write,

$$5T + 2s = 6T$$

By taking $5T$ from both sides of the above equation it becomes,

$$2s = T \text{ or } s = \frac{T}{2} \qquad \text{(iv)}$$

and so on.

Change of Reference Note

If the reference note of a set is required to be changed, then this can be done, without in any way changing the ratios previously existing between the members of the set, by simply dividing through by that ratio corresponding to the new reference note, or what amounts to the same thing, multiplying each member of the set by the reciprocal of this ratio. This has been done for D ($\frac{9}{8}$) in set (b) of this chapter—the result of which is given as set (e) of the same section. In this same way, the reference note C ($\frac{1}{1}$) of set (b) could be changed to G ($\frac{3}{2}$) by multiplying all its ratios by $\frac{3}{2}$.

In terms of logarithmic units, the above operation is carried out by subtracting the logarithmic measure (say, in cents) of the interval made by the new reference note with the old, from all the numbers in the set—after having first added on 1200 to those whose value is less than this. While in terms of deviation from a chosen standard set—as in Table A, pp. 242-3—the same operation can be carried out by reducing the new reference note's deviation to zero, and then adding (or subtracting) the same deviation to (or from) all the others of the set.

The Use of the Deviation Ratios and Row J-V of Table A to Obtain the Frequencies of the Remaining Temperaments

This method has the advantage that only one of the two figures involved in the calculation is approximate—that of the deviation—the frequencies of row J-V being *exact* for an A based on 440 cycles per second.

Suppose, as an example, that the frequencies of Basic Mean-Tone are required to be derived in this manner. Then $k/4$, a's deviation in this temperament, must first be subtracted from all the deviations given in row M-1. When this is done it can be seen that D♭, F, and A will each have zero deviation and will therefore have exactly the same frequencies as their counterparts in row J-V. The intervals made by A♭, C, E and G♯ will each have deviations of $-k/4$, which, in terms of a decimal ratio, is ·9969.

Hence the new frequencies corresponding to

A♭ will be ·9969 × 422·4 = 421·09 cycles per second
C ,, ,, ,, × 264·0 = 263·18 ,, ,, ,,
E ,, ,, ,, × 330·0 = 328·98 ,, ,, ,,
G♯ ,, ,, ,, × 412·5 = 411·22 ,, ,, ,,

and so on for the remaining intervals with deviations of $-k/2$ and $-3k/4$.

21

SUMMARY OF MATHEMATICAL TERMS AND OPERATIONS

Sum—addition—multiplication—product—factor—prime factor—prime number—multiple—continued product—involution—power—index or exponent—laws of indices—base—logarithm—decimal system—logarithm to the base ten—ratio—logarithmic scale—difference—subtraction—division—quotient—remainder—mutually inverse operations—equation—transforming—evolution—root—fraction—vulgar fraction—denominator—numerator—decimal fraction—reciprocal—significant figures—recurring decimal fractions—reduction—lowest terms—rules of divisibility—improper fractions—mixed numbers—comparison of fractions—important fundamental theorems on fractions—addition and subtraction of fractions—multiplication and divison of fractions—reciprocals of fractions—involution and evolution of fractions—continued fractions—series—term—arithmetic series—geometric series—harmonic series—direct and inverse variation or proportion—graph or mathematical curve.

The operation performed in obtaining the *sum* of two (or more) numbers is called *addition*. The sign of addition is $+$ ('plus').

Thus $4 + 3 = 7$, the sign '$=$' ('equals' or 'is the same as') being here used to connect the operating instructions with the result of carrying them out.

The operation performed in obtaining the sum of a given number of repetitions of the same number—a process of continued addition—is called *multiplication*, and the resulting sum is called the *product* of the number and its number of repetitions. The sign of multiplication is $\times$ ('multiplied by' or 'times').

Thus $3 \times 4 = 4 + 4 + 4 = 12$.

The sign '$\times$' is usually omitted in multiplications involving numbers and quantities represented by letters, or separated by brackets. Thus,

$2a$ means $2 \times a$

$3(T - t)$ means $3 \times (T - t)$ and equals $3T - 3t$

The numbers multiplied together to give the product are sometimes known as *factors* of the product. Thus 6 and 4 are *factors* of 24, so also are 2 and 3, since $2 \times 2 \times 2 \times 3 = 24$. The latter are known as *prime factors* since they are all prime numbers.

A *prime number* is a number which has no factors except itself and one.

Any given set of numbers having no common factor between them are said to be *prime to one another.* They need not be prime numbers. For example, the numbers 4, 9, and 25 are prime to one another.

A number is said to be a *multiple* of any one of its factors, since it contains any one of them an exact number of times without remainder.

A product involving more than two factors is sometimes called a *continued product.*

Thus, 60 is the continued product of 3, 4 and 5.

The operation performed in obtaining the continued product of a number with itself is called *involution*, and the resulting continued product is known as a *power* of the number. The small figure placed at the right-hand top corner to indicate how many factors are involved in the product is called an *index* (plural *indices*), or *exponent.*

Thus, the first power of three, $3^1 = 3$

the second power (or 'square') of three, $3^2 = 3 \times 3 = 9$

the third power (or 'cube') of three, $3^3 = 3 \times 3 \times 3 = 27$

the fourth power of three, $3^4 = 3 \times 3 \times 3 \times 3 = 81$

etc.

The product of two or more powers of the same number (or quantity) may be obtained by adding the indices of the powers.

Thus $2^2 \times 2^3 = 2^{2+3} = 2^5$ (*i.e.* $4 \times 8 = 32$)

Also, in the same manner, their quotient may be obtained by subtracting the indices. Thus:—

$2^5 \div 2^2$ or $\frac{2^5}{2^2} = 2^{5-2} = 2^3$ (*i.e.* $32 \div 4 = 8$)

And $(2^2)^3 = 2^2 \times 2^2 \times 2^2 = 2^{2\times3} = 2^6$ (*i.e.* $4 \times 4 \times 4 = 64$)

These three statements, when generalised as follows,

(1) $a^m \times a^n = a^{m+n}$

(2) $a^m \div a^n$ or $\frac{a^m}{a^n} = a^{m-n}$ (invalid for $a = 0$)

(3) $(a^m)^n = a^{mn}$

are known as the *laws of indices.*

Three important results of these laws are listed below.

(1) $a^n \div a^n$ or $\frac{a^n}{a^n} = a^{n-n} = a^0$

But any number divided by itself must equal one. It therefore follows that *any number (or quantity) raised to the power* 0 *must equal* 1. As will be appreciated later, this means that the logarithm of 1 to any base $= 0$.

(2) $a^2 \div a^4$ or $\frac{a^2}{a^4} = a^{2-4} = a^{-2}$, but $\frac{a^2}{a^4}$ also equals $\frac{1}{a^2}$ therefore

$a^{-2} = \frac{1}{a^2}$ and generally $a^{-n} = \frac{1}{a^n}$

(3) $a^{\frac{1}{2}} \times a^{\frac{1}{2}} = a^{\frac{1}{2}+\frac{1}{2}} = a^1$ or a, but $\sqrt{a} \times \sqrt{a}$ also equal a. It follows therefore that $a^{\frac{1}{2}}$ and $\sqrt{a}$ are simply two different ways of denoting the same thing—and so on.

Thus, *generally,* $a^{\frac{1}{n}}$ *equals the n-th root of a, or* $\sqrt[n]{a}$.

Also $a^{\frac{m}{n}} = (a^m)^{\frac{1}{n}} = \sqrt[n]{a^m}$ or $(a^{\frac{1}{n}})^m = (\sqrt[n]{a})^m$

e.g. $8^{\frac{2}{3}} = (8^2)^{\frac{1}{3}} = \sqrt[3]{8^2} = \sqrt[3]{64} = 4$

or $= (8^{\frac{1}{3}})^2 = (\sqrt[3]{8})^2 = 2^2 = 4$

The number (or quantity) which is raised to the power, such as 3 or 8 (or *a*) in the above examples, is called the *base.*

Note that all roots and powers of one equal one.

Note that all powers and roots of 1 equal 1.

The introduction of such fractional and negative indices makes it possible to express any number—except the number 1 —as a power of any other number.

e.g. $4 = 2^2 = 3^{1.2619} = 4^1 \ldots = 10^{0.6021} \ldots = 16^{0.5} \ldots$

$\frac{1}{4} = 2^{-2} = 3^{-1.2619} = 4^{-1} \ldots = 10^{-0.6021} \ldots = 16^{-0.5} \ldots$

etc.

Whenever numbers are so expressed—i.e., as a power of a chosen single number (called the *base*)—the indices of these powers can be referred to as the *logarithms* of the numbers to that base.

Thus the equation

$$8 = 2^3 \text{ (read `eight equals two cubed')}$$

may be also written as

$$\log_2 8 = 3 \text{ (read `the logarithm of eight, to the base two, equals three')}.$$

The advantage of expressing numbers in this way, *i.e.*, as powers of a common base, is that by making use of the laws of indices and a table of logarithms (*i.e.* indices of the base):

(1) The operation of *multiplication* can be reduced to one of *addition* of indices.

(2) The operation of *division* (defined later) can be reduced to one of *subtraction* of indices.

(3) The operation of *involution* can be reduced to one of *multiplication* of indices.

(4) The operation of *evolution* (defined later) can be reduced to one of *division* of indices.

Our system of numbers is a *decimal system* based on powers of ten as indicated by the place or position of the figures denoting them.

Thus

$$659{\cdot}42 = 6 \times 10^2 + 5 \times 10^1 + 9 \times 10^0 + 4 \times 10^{-1} + 2 \times 10^{-2}$$

Here the powers of ten may be regarded as 'decimal point shifting operators', the sign and number of whose index gives the direction and number of places the point is required to be shifted—a positive index indicating a shift to the left, a negative index, a shift to the right and a zero index, no shift at all.

Hence, by employing a suitable power of ten as a multiplier, any number can, in effect, be made to lie between 1 and 10, *i.e.*, to correspond to a fractional power of ten having an index lying between 0 and 1. Tables of such numbers and indices have been compiled and are known as *logarithms to the base* 10, or *common logarithms*. With the aid of such a set of tables any number can be expressed as a power of ten.

For example,

$$625{\cdot}4 = 6{\cdot}254 \times 10^2 = \text{(using the table) } 10^{\cdot 7962} \times 10^2$$
$$= 10^{2+\cdot 7962} = 10^{2\cdot 7962}$$

$$\cdot 0006254 = 6\cdot 254 \times 10^{-4} = \text{(using the table) } 10^{\cdot 7962} \times 10^{-4}$$
$$= 10^{\cdot 7962-4}$$

which is written as $10^{\bar{4}\cdot 7962}$ so as to keep the fractional part positive (logarithm tables do not cater for negative values).

It is common practice to omit the '10's' and write, for example, 2·7962 equals the logarithm of 625·4, but care must be taken to keep numbers and logarithms separate.

The number corresponding to a given logarithm may be obtained by reversing the above process—using a table of antilogarithms, if preferred.

Readers who may have wondered where the 4000 (or 3986) in the equation for obtaining the number of cents in a given ratio (p. 238) comes from—or why there are 301·03 savarts in an 'octave'—should now be in a better position to appreciate the following explanation.

The logarithm of a ratio gives a direct measure of it (see later), and if the standard unit of ratio chosen is $\frac{2}{1}$ (*i.e.* an 'octave') then the logarithms used must necessarily be to the base 2.

In such units the value (say, I) of any ratio 'r' would be given by

$$I = \log_2 r \text{ 'two-to-ones' or 'octaves'}.$$

For ratios which are exact powers of 2 (*i.e.* which correspond to whole numbers of octaves) this formula is, of course, easy to use—the number of 'octaves' being equal to the index of that power of 2 corresponding to the ratio. For example, since $2^4 = 16$, then the ratio $\frac{16}{1}$ corresponds to 4 octaves, or in terms of the formula above, equals

$$I = \log_2 r = \log_2 2^4 = 4 \text{ 'two-to-ones' or 'octaves'}.$$

So far as other ratios are concerned—although logarithms to the base 2 are not generally available, fortunately, logarithms to the base 10 are, and the logarithm to the base 2 of any number equals the logarithm to the base 10 of that number divided by ·3010. (Readers who wish to check up on this can find it in any good book on algebra under 'change of base of logarithms'.) Thus the formula above may be rewritten as

$$I = \log_2 r = \frac{\log_{10} r}{\cdot 3010} \text{ 'two-to-ones' or 'octaves'}.$$

If, as in the case of the cent, the 'octave' is divided into 1200 equal logarithmic units (*i.e.* 100 for each semitone of equal temperament) then the above formula becomes

$$I = \frac{1200}{\cdot 3010} \log_{10} r = 3986 \log_{10} r \text{ or (approx.) } 4000 \log_{10} r \textit{ cents.}$$

If, on the other hand, as in the case of the savart, the 'octave' is divided into 301 equal logarithmic units, then the formula becomes

$$I = \frac{301}{\cdot 3010} \log_{10} r = 1000 \log_{10} r \textit{ savarts.}$$

(·30103 is a more accurate version of ·3010.)

Suppose that it is required to compare the lengths of two tables and that no graduated scale or rule is available. One way of doing this would be to obtain the length of each table in terms of 'hands'. If, after doing this, one table was found to be 20 hands in length and the other 10, then the longer table would be said to be twice as long as the shorter one. This latter statement is independent of the 'unit' of length used (in this case the 'hand') for the same conclusion would have been reached had the unit in which the tables were measured been a foot, the length of the shorter table, the length of the longer table, or in fact any length whatever—although obviously, from the point of view of accuracy and simplicity, the most convenient unit of length to use in any particular case would be the largest unit which was contained exactly in each of the lengths being compared. In the simple case considered above the comparison is sometimes expressed by saying that the table lengths are in the *ratio* two to one, written 2 : 1, or, as is done throughout this book, in the form of a fraction—as $\frac{2}{1}$.

Ratios of $\frac{0}{1}$, $\frac{1}{0}$, or $\frac{0}{0}$ have no practical meaning.

The idea of ratio is a very fundamental one and springs from the conception of geometrical similarity. Its importance in musical acoustics lies in the two experimentally derived facts stated below.

(1) Any pair of notes—whatever their pitches—generated in the ear by vibrations having the same frequency ratio are accepted by the musician as having a similar musical relation to one another, *i.e.* are recognised as the same musical interval. (Equal ratios in intensity are also perceived as approximately equal changes in loudness.)

(2) The resultant frequency ratio obtained by multiplying together the frequency ratios corresponding to any two, or more, intervals, is found to be that ratio corresponding to the interval which is judged by the ear to be their sum.

Similarly, dividing one frequency ratio by another is found, in effect, to subtract the intervals to which the frequency ratios correspond.

It follows from (1) that intervals can be specifically defined, fixed, or *physically standardised*, by their corresponding frequency ratios, and that identical frequency ratios may be expected to produce identical intervals in the ear.

Fact (2) tells us that intervals can be added or subtracted *physically* by simply multiplying or dividing their corresponding frequency ratios.

But before proceeding with this subject it is worth considering the following purely mathematical points.

The comparison of two quantities of the same kind can be made and expressed in two different ways.

(1) The *amount* by which one quantity exceeds, or is greater than another, is obtained by subtracting the smaller from the greater and hence finding their *difference*—or the *remainder*. Since this difference can be anything from a fraction of the smaller to many times its size, the comparison is often more conveniently and significantly expressed—especially in the latter case—in terms of the *ratio* of the two quantities.

(2) The number of *times* that one quantity is greater than another is obtained by their *ratio*, *i.e.* by dividing the greater by the smaller and hence finding how many times—including any fraction of a time—the smaller quantity has to be subtracted from the larger so as to leave *no remainder*.

Likewise, the comparison of ratios can also be made and expressed in two different ways.

(1) The *amount* by which one ratio exceeds, or is greater than another, is obtained by *dividing* one ratio by another, *i.e.* by finding out how many times—including any fraction of a time—the smaller ratio has to be subtracted from the larger so as to leave *no remainder*.

(2) The number of *times* one ratio is greater than another is found by obtaining how many times—including any fractions of a time—the *continued product* of the smaller is contained in the

larger, or, to what *power*—including any fraction of a power—the smaller has to be raised so as to equal the larger. Alternatively, by finding how many times—including any fraction of a time—the larger needs to be *continually divided* by the smaller so as to reduce it to unity ratio, $\frac{1}{1}$. Either process involves, in effect, finding the difference between the logarithms of each ratio, or what amounts to the same thing, finding the logarithm of the ratio of the two ratios.

Returning now to the subject of intervals—from all that has been said above it follows that if two or more identical intervals are to be added physically, then the ratio defining the resulting interval must equal the continued product of the ratio of any one of the intervals with itself, repeated two or more times as required, *i.e.*, must equal that ratio which corresponds to one of the component intervals raised to the power n, where n is the number of identical intervals whose sum is required.

For instance, the ratio corresponding to the sum of three major thirds would equal

$\frac{5}{4} \times \frac{5}{4} \times \frac{5}{4} = (\frac{5}{4})^3 = \frac{125}{64}$. This is less than an octave ($\frac{128}{64}$) by the ratio

$\frac{2}{1} \div \frac{125}{64} = \frac{2}{1} \times \frac{64}{125} = \frac{128}{125}$ (a *diesis* or 42 cents).

Conversely, if it is required to divide any interval into say, n equal intervals, then the ratio corresponding to each of the submultiple intervals must equal the n-th root of the ratio corresponding to the interval being divided.

Thus to divide an octave into 12 equal intervals, as is done in equal temperament, it is necessary to obtain the 12-th root of $\frac{2}{1}$ or 2 (any root or power of $1 = 1$). Now $\sqrt[12]{2} =$ antilog $\frac{\log 2}{12} =$ antilog $\frac{\cdot 3010}{12} =$ antilog $\cdot 0251 = 1\cdot 059$.

Therefore the ratio corresponding to $\frac{1}{12}$ of an octave is $\frac{1\cdot 059}{1}$.

This result was obtained using four-figure logarithmic tables. If a more accurate result is required seven-figure tables should be used.

As previously indicated, no matter what place a figure occupies, moving it 1, 2, etc. places to the left (or its decimal point 1, 2, etc. to the right) has the effect of multiplying it by 10 (*i.e.* 10^1), 10^2, etc. That is to say, so far as position is concerned, equal amounts of shift from right to left by any whole number of places multiplies the number, whatever its initial position, by that power of ten equal to the number of places shifted.

Now suppose it is required to extend this method of multiplication to include multiplication by numbers less than ten. In this case it is obvious that the amount of shift required would need to be some fraction of one place. Following this up, imagine, for the sake of simplicity, the distance between adjacent figures (*i.e.* the 'length' of a place) to be increased to, say, 10 inches; what then would be the shift required for, say, '$\times 3$'? Since 3×3 or $3^2 = 9$, it follows that two of these '$\times 3$' shifts would have to add up to a little less than the length of one place (10 inches in our case) so that one would have to be somewhat less than half the length of a place (5 inches in our case). This fact, in turn, suggests the question 'What number when multiplied by itself (*i.e.* squared) equals 10 exactly?'—the answer to which is $10^{\frac{1}{2}}$ or $\sqrt{10}$ or 3·162. Thus a shift of ½ a place must be exactly equivalent to '$\times 10^{\frac{1}{2}}$' (or '$\times\sqrt{10}$' or '$\times 3{\cdot}162$'). In the same way a shift of ⅓ of a place would be equivalent to a '$\times 10^{\frac{1}{3}}$' (or '$\times\sqrt[3]{10}$' or '$\times 2{\cdot}154$')—and so on.

The important point to note here is that the fraction of a place required to bring about multiplication by a given number equals the index of the fractional power of ten corresponding to that number; *i.e.*, the *logarithm* of that number. What has in fact been described is a *logarithmic scale* such as can be seen on any slide-rule (in our case, the 1 to 10 scale of a ten-inch slide-rule).

So far as musical acoustics is concerned the usefulness of the logarithmic scale lies in the fact that the logarithm of a ratio is directly proportional to the ratio itself, and thus provides a means of measuring ratios directly. For instance, if the logarithm of one ratio is twice that of the logarithm of a second ratio, then the first ratio is twice the size of the second and can be represented graphically by lengths in the ratio of 2 to 1—and so on. But since the width of intervals is directly proportional to such measures of their defining frequency ratios, it follows that intervals can also be measured and represented in the same way by the logarithms of their corresponding frequency ratios, or by lengths proportional to them. In such a scale, the same length represents the same ratio—and therefore its corresponding interval, whatever its position in pitch. Logarithmic scales are used in this book wherever intervals are represented graphically.

It is interesting to note that the word 'logarithm' is derived from the Greek, *logos*—meaning ratio, and *arithmos*—meaning number (count or measure).

In developing the above ideas use has been made of terms not previously defined, such as subtraction, division, root, etc. These matters will now be dealt with.

The operation performed in obtaining the *difference* of two numbers is called *subtraction*. The sign of subtraction is — ('minus').

$$\text{Thus } 5 - 3 = 2.$$

The operation performed in finding how often one number is contained in another—a process of continued subtraction—is called *division*, and the number of such subtractions possible is called the *quotient*. There are several signs for division, ÷, /, and the horizontal line between the two numbers forming a fraction.

Thus $12 \div 3$ or $\frac{12}{3} = 4$, because 3 can be taken away from 12 exactly 4 times. If the number of subtractions possible cannot be expressed exactly as a whole number, then the amount left over—after the highest possible number of complete subtractions have been made—is called the *remainder*.

The operation of addition—and therefore also of multiplication—can be performed in any order. This is not so for subtraction, nor therefore for division.

Thus $5 - 2$ is not the same as $2 - 5$, nor is $2 \div 5$ the same as $5 \div 2$.

Addition and subtraction are *mutually inverse operations* in that, if the same amount is both added and subtracted from a given number, that given number remains unaltered. Thus, as before, multiplication and division must also be mutually inverse operations.

Any group of quantities connected by an 'equals' sign forms what is known as an *equation*.

The related quantities of any question can be expressed in more than one form, and it is important to be able to transpose the equation to that form which best suits the particular problem in hand. The simplest way of changing the form of an equation, or *transforming* it, is to make use of the axioms stated immediately below to introduce or remove any particular quantity to or from either side of the equation.

Quantities which are equal remain equal when the same quantity is added to or taken from them and when they are multiplied or divided by the same quantity.

The operation performed in obtaining a number from a given power of that number is called *evolution*, and the resulting number is called the *root* or base of the power of the number.

From this definition it follows that involution and evolution must also be mutually inverse operations.

Thus, since $144 = 12^2$, therefore the square (second) root of $144 = 12$, *i.e.* $\sqrt{144}$ or $144^{\frac{1}{2}} = 12$.

Also, since $343 = 7^3$, therefore the cube (third) root of $343 = 7$, *i.e.* $\sqrt[3]{343}$ or $343^{\frac{1}{3}} = 7$.

And, since $2 = 1{\cdot}0594\ldots^{12}$, therefore the twelfth root of $2 = 1{\cdot}0594\ldots$, *i.e.* $\sqrt[12]{2}$ or $2^{\frac{1}{12}} = 1{\cdot}0594\ldots$

and so on.

Ratios of two quantities are very often expressed in the form of fractions, in which case they may be dealt with using the standard techniques used in the treatment of fractions. The following notes on fractions are given with this point in view.

If the whole of 'something' is divided into, say, five equal parts, then each part would equal 'one divided by five', *i.e.* ⅕ or one-fifth of the whole 'something', and is known as a *fraction*—or more precisely—a *vulgar fraction* (vulgus—'common', fractus—'broken'). The '5' or 'bottom' of the fraction is called the *denominator* because it is from this part that the fraction takes its name. Three such parts are written as ⅗ where '3', the 'top' of the fraction, is called the *numerator* since it denotes the number of fifths. There is another way of regarding the fraction $\frac{3}{5}$. It may also be considered as three whole 'somethings' divided (*i.e.* shared equally) by five.

By virtue of our decimal system of notation, vulgar fractions whose denominators are 10, 100, 1000, etc. may, by shifting their numerator's decimal point an appropriate number of places to the left, be written with their denominators omitted (*i.e.* '1' understood). Any vulgar fraction may be expressed in this form by dividing its numerator by its denominator. Such fractions are called *decimal fractions*.

e.g. $\frac{32}{25} = \frac{32 \times 4}{25 \times 4} = \frac{128}{100} = \frac{1 \cdot 28}{1 \cdot 00} = 1 \cdot 28$

Alternatively, dividing both top and bottom of the fraction by 25

$$\frac{32}{25} = \frac{\frac{32}{25}}{\frac{25}{25}} = \frac{32 \div 25}{1} = \frac{1 \cdot 28}{1} = 1 \cdot 28$$

or, more simply,

$$\frac{32}{25} = 32 \div 25 = 1 \cdot 28$$

To change a decimal fraction to its equivalent vulgar fraction form requires the above process to be applied in reverse.

e.g. $1 \cdot 125 = \frac{1 \cdot 125}{1 \cdot 000} = \frac{1125}{1000} = \frac{225}{200} = \frac{45}{40} = \frac{9}{8}$

The *reciprocal* of any number equals one, divided by that number and is often quoted in its decimal fraction form. The operation of division can be replaced by that of multiplication —and vice-versa—by the use of such reciprocals.

e.g. $\frac{32}{25} = 32 \times \frac{1}{25} = 32 \times 0 \cdot 04 = 1 \cdot 28$

A table of reciprocals is given at the end of the book.

No operation involving measurement—except that of simple counting—can ever be carried out *exactly*. The resultant accuracy will depend on the sensitivity and accuracy of the instruments used, the skill of the operator, and innumerable other factors. For this reason no measurement, or calculation involving figures obtained by measurement, should ever be stated to a greater degree of accuracy, *i.e.*, to more *significant figures* or figures of importance, than can be justified by the data and method used to obtain it.

A length written as ·024 inch is said to be given to two

significant figures. If the length were in fact ·0246 inch, then, since 46 is nearer to 50 than to 40, its correct length to two significant figures should be stated as ·025 inch, and so on.

Again, consider the following vulgar fractions,

$$\frac{64}{45} = 64 \div 45 = 1{\cdot}42222\ldots$$
$$\frac{128}{135} = 128 \div 135 = 0{\cdot}9481\ 9481\ 9481\ldots$$

These are known as *recurring decimal fractions* and are denoted as $1{\cdot}4\dot{2}$ and $0{\cdot}\dot{9}48\dot{1}$ respectively. Since their recurrence has no end it is obviously impossible to write them down completely, and therefore exactly. However in any practical calculation it is only necessary to write down enough of their decimal places as is sufficient to match the accuracy of the other quantities used with them.

The value of a fraction is not altered by the multiplication or division of both its numerator and denominator by the same number. The first of these processes is often used in the comparison, addition and subtraction of fractions, while the latter process, known as *reduction*, is used to reduce a fraction to its lowest terms. A fraction is said to be in its *lowest terms* when no factor common to both numerator and denominator remains. Such common factors are best found by first applying the *rules of divisibility*, quoted below, and then testing directly for divisibility by all the remaining prime numbers of value up to that of the square root of the numerator and denominator respectively.

A number is divisible by

2 if its last figure is even.
3 if the sum of its figures is divisible by 3.
5 if its last figure is 0 or 5.
7 no simple rule exists—test by direct division by 7.
11 if the sum of one of its sets of alternate figures differs from the sum of the other set by 0 or some multiple of 11. Thus the number 179,135 is divisible by 11 since the difference between $(1 + 9 + 3), = 13$, and $(7 + 1 + 5), = 13$, is zero.

If a certain number is divisible by two (or more) prime numbers it is also divisible by the product(s) of these prime numbers.

Since all quantities may be assumed to have an 'understood'

denominator of 1, it follows that quantities greater than 1 can be expressed in the form of a fraction. Any such fraction is known as an *improper fraction*—since its numerator is always greater than its denominator.

Taking, for example, the following *mixed numbers*, then,

$$2\,^2/_3 = {}^2/_1 + {}^2/_3 = {}^6/_3 + {}^2/_3 = {}^8/_3$$
$$1\,^1/_5 = {}^1/_1 + {}^1/_5 = {}^5/_5 + {}^1/_5 = {}^6/_5$$

The multiplication of such numbers is generally made much simpler by first expressing them as improper fractions—in which form most of the ratios in this book are expressed.

Fractions may be compared by making either their numerators or their denominators equal—as shown on p. 209.

Two important theorems on fractions are stated in Chapter 19. These can be deduced directly from the fact already mentioned —that multiplication or division of both numerator and denominator by the same quantity leaves the value of the fraction unaltered—for this is the same as adding (or subtracting) a certain fraction or percentage of the numerator to (or from) itself, and at the same time, adding (or subtracting) the *same fraction* or percentage of the denominator to (or from) itself. Whereas, adding or subtracting the *same quantity* to or from both numerator and denominator must change the fraction as indicated in Chapter 19. These two operations are in fact particular cases of the general principle given in Appendix 9.

Addition and subtraction of fractions having the same denominators can be carried out simply by the addition or subtraction of their numerators. If the denominators are different, then these must first be made the same by the multiplication of both numerator and denominator of each fraction by the lowest number which will make all the denominators equal.

The result of the *multiplication of* two (or more) *fractions* is that fraction whose numerator is the product of the separate numerators and whose denominator is the product of the separate denominators.

The *division of one fraction by another* is carried out by multiplying the first fraction by the reciprocal of the second (*i.e.* the one that comes after the division sign). The *reciprocal* of any fraction is simply the fraction inverted. Thus $5/4$ and $4/5$ are reciprocals.

The *involution* (raising to the power) *of fractions* is brought about by raising both numerator and denominator to the required power.

Thus $(5/4)^2 = 5^2/4^2 = 25/16$ and $(9/8)^3 = 9^3/8^3 = 729/512$.

The *evolution* (finding of the root) *of fractions* is carried out by obtaining the root of both numerator and denominator.

Thus $\sqrt[3]{27/8} = \sqrt[3]{27}/\sqrt[3]{8} = 3/2$ and $\sqrt{5/4} = \sqrt{5}/\sqrt{4} = \sqrt{5}/2$.

Continued fractions are dealt with in Appendix 9.

A mathematical *expression* is a collection of one or more signs, figures, and/or letters, which are used to denote one quantity. Thus, in the equation,

$$5th = 2T + t + s$$

the '$2T + t + s$' is an expression denoting the single quantity '5th' or 'one fifth'. The component parts of such an expression —*i.e.*, those parts separated by '+' or '−' signs—are known as the *terms* of the expression.

Terms forming a single expression are put in brackets wherever and whenever it is essential to indicate or emphasise this fact. For example,

$$\begin{aligned} 4(5th) &= 4(2T + t + s) \\ &= 8T + 4t + 4s \end{aligned}$$

The horizontal line between the numerator and denominator of a fraction is also virtually a bracket.

The rules governing the removal of brackets can be deduced by considering the following numerical examples:—

$6 + (4 + 1 - 2)$	means	$6 + 3$	or	$6 + 4 + 1 - 2 = 9$
$6 - (4 + 1 - 2)$	,,	$6 - 3$	,,	$6 - 4 - 1 + 2 = 3$
$6 - (4 - 1 + 2)$	,,	$6 - 5$	,,	$6 - 4 + 1 - 2 = 1$
$6(4 + 1 - 2)$	,,	6×3	,,	$24 + 6 - 12 = 18$
$-6(4 + 1 - 2)$	,,	-6×3	,,	$-24 - 6 + 12 = -18$

etc.

From these it can be seen that, while brackets preceded by a positive sign or quantity can be removed without changing any of the signs of the terms enclosed in the bracket, those preceded by a negative sign or quantity can only be removed provided that each sign within the bracket is reversed. Note that a quantity without a sign preceding it is understood to be a positive quantity.

In forming an equation it is sometimes necessary to make use of brackets within brackets. In simplifying such an equation it is usually thought best to remove the innermost bracket first.

As an example of this kind consider the following question. By how much—in terms of diatonic steps—do four fifths exceed the sum of two octaves and a major third?

$$4(5\text{th}) - [2(8\text{th}) + M3]$$
$$= 4(2T + t + s) - [2(3T + 2t + 2s) + (T + t)]$$
$$= 8T + 4t + 4s - [6T + 4t + 4s + T + t]$$
$$= 8T + 4t + 4s - [7T + 5t + 4s]$$

see eqs. (k), (m), (p), of Chapter 19.

$$= 8T + 4t + 4s - 7T - 5t - 4s$$
$$= T - t \text{ or } k$$

from eq. (qa) of Chapter 19.

In order to avoid any possible ambiguities in carrying out the simplification of a complex expression—*i.e.*, one in which several different kinds of operations are required to be performed—the following order of priority has been generally agreed upon,

(i) Brackets
(ii) Division
(iii) Multiplication
(iv) Addition
(v) Subtraction

Note that the long horizontal line between the numerator and denominator of a fraction not only denotes division but also implies brackets. The horizontal line sometimes used to extend a root sign also implies a bracket.

Thus, the complex expression,

$$5 - 2 + \frac{3\times4-2+6}{5+8\times3-21}$$

may be rewritten as

$$5 - 2 + (3 \times 4 - 2 + 6) \div (5 + 8 \times 3 - 21)$$
$$= 5 - 2 + (12 - 2 + 6) \div (5 + 24 - 21)$$
$$= 5 - 2 + (18 - 2) \div (29 - 21)$$
$$= 5 - 2 + 16 \div 8$$
$$= 5 - 2 + 2$$
$$= 7 - 2$$
$$= 5$$

Here, though brackets come first, they cannot be removed until the operations enclosed by them have been completed in accordance with the remaining order of priority. When this has been done the same order of priority is again applied, giving the answer, 5.

A *series* is a succession of numbers (or quantities) arranged in order according to some definite law. Each number is called a *term* of the series.

If the successive terms of a series increase (or decrease) by a *common difference* they are said to form an *arithmetic series.*

E.g. the series 3, 7, 11, 15, etc.—which has a common difference of 4.

If the successive terms of a series increase (or decrease) by a *common ratio* they are said to form a *geometric series.*

E.g. the series 3, 6, 12, 24, etc.—which has a common ratio of 2.

A succession of quantities are said to form a *harmonic series* when their reciprocals form an arithmetic series.

The ratios of the periods of the partial vibrations of any complex periodic vibration form the harmonic series

$$\frac{1}{1} \quad \frac{1}{2} \quad \frac{1}{3} \quad \frac{1}{4} \quad \frac{1}{5} \quad \frac{1}{6} \quad \text{etc.}$$

Whereas the reciprocals of these ratios give the corresponding frequency ratios

$$\frac{1}{1} \quad \frac{2}{1} \quad \frac{3}{1} \quad \frac{4}{1} \quad \frac{5}{1} \quad \frac{6}{1} \quad \text{etc.}$$

—an arithmetic series.

Any succession of ratios corresponding to a succession of identical intervals must form a geometric series

e.g. in 'octaves' $\frac{1}{1} \quad \frac{2}{1} \quad \frac{4}{1} \quad \frac{8}{1}$ etc.

in 'fifths' $\frac{1}{1} \quad \frac{3}{2} \quad \frac{9}{4} \quad \frac{27}{16}$ etc.

Adjacent terms of any series are said to be 'in arithmetic, geometric, harmonic, etc., *proportion*'.

Whenever two quantities are so related that an *increase* in one brings about an *increase* in the other, then either quantity is said to *vary directly* as, or to be *directly proportional* to, the other. On the other hand, if an *increase* in one causes a *decrease* in the other then either quantity is said to *vary inversely* as, or to be *inversely proportional* to, the other.

The distance covered in moving for a given time at any given speed is directly proportional to both time and speed—either doubling the time or the speed doubles the distance covered. Whereas, in covering a given distance, the time required varies inversely as the speed—*i.e.* if the speed is doubled the time is halved, and vice-versa.

Again, the frequency of a perfectly flexible and uniform string, subjected to a constant tension, varies inversely as its length. Thus to double its frequency the length must be halved; to halve its frequency the length must be doubled; to increase its frequency in the ratio $^3/_2$ the length must be reduced to $^2/_3$ of its original length—and so on.

A *graph* or *mathematical curve* is a line drawn in such a way as to show pictorially how the value of one quantity is related to the value of another, *e.g.*, the sound-curves on p. 26, which relate time and displacement.

Usually the values of the two quantities concerned are made proportional to the distance from two straight lines at right angles to each other (the 'axes of reference'). Assuming this to be so, then only if one quantity varies directly as the first power of another will the line relating them be straight, *e.g.*, the length and weight of a uniform wire, but *not* the radius and weight of a solid sphere. For this reason the type of relationship between the length and weight of a uniform wire is sometimes said to be *linear*. Relationships which are not linear are then said to be *non-linear*.

The conception of the physical terms cycle, period, frequency, amplitude, phase, simple and complex form, harmonic content etc., of a vibration, and the difference between a periodic and a non-periodic vibration, are most easily and quickly grasped when displacement and time are presented simultaneously in graphical form.

The graph shown opposite is a representation (an associated curve) of simple harmonic motion occurring in time and/or (air) space. Variations in both time and space (or length) may be assumed to be plotted horizontally, but to avoid having to reproduce the same diagram twice, time is indicated below the graph and space above it. Displacement may be assumed to be plotted vertically.

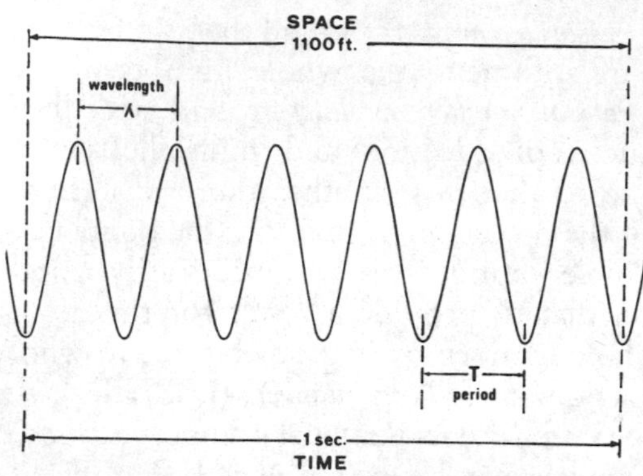

In this particular graph it is easy to see that there are 6 complete cycles in a length of 1100 ft. Thus it follows that the length of one 'wave' or cycle—*i.e.*, the wavelength λ)—will be 1100 ft. divided by 6 which equals $183{\cdot}\dot{3}$ ft. Likewise, in the time of 1 second, 6 cycles are completed; therefore the time for one cycle—the periodic time (T)—must equal $\frac{1}{6}$ of a second (per cycle). Another way of looking at this is to say that there are 6 cycles each second, *i.e.*, the frequency (f) is 6 cycles per second (or hertz)—which, as a note, lies below our audio range.

Suppose now, for instance, that both time and distance were reduced to $\frac{1}{10}$ of their values—leaving the diagram otherwise unaltered—then the SPACE would be 110 ft., the TIME $\frac{1}{10}$ of a second, the wavelength $18{\cdot}3$ ft., and the frequency 60 cycles per second, *i.e.*, about two octaves and a tone below middle C.

Notice that, in both the above calculations, the ratio SPACE (or distance) to TIME gives the speed of sound in air, *i.e.*, $^{1100}/_{1}$ equals $^{110}/_{0}$ 1 equals 1100 feet per second. The same condition must be satisfied for any similar calculation to be valid.

In term of formulae,

$$f = \frac{1}{T} \quad \text{or} \quad T = \frac{1}{f}$$
$$\text{and } \lambda = \frac{1100}{f} \quad \text{or} \quad f = \frac{1100}{\lambda}$$

For metrical working the '1100 feet' should be replaced by '332 metres'.

The shape of a corresponding pressure variation graph would be virtually identical, though, strictly speaking, the whole curve would be (phase) shifted bodily to the right by $\frac{1}{4}$ of a cycle.

The real system of numbers. Of all the numbers considered in the study of arithmetic the whole numbers 1, 2, 3, . . . are called the *natural numbers* or *integers*, and with these numbers the operations of addition and multiplication are always possible, the result being another number. However when considering the operation of division, the quotient of any two positive whole numbers is not necessarily another whole number; so that to provide a result for the division of any positive whole number by any other positive whole number, positive fractions or *rational numbers* (*i.e.*, ratios), such as $\frac{2}{3}$, $\frac{5}{7}$ *etc.*, must be introduced. Rational numbers include the natural numbers—which may be expressed as $\frac{1}{1}$, $\frac{2}{1}$, $\frac{3}{1}$, etc.

In a similar way the operation of subtraction is not possible with all positive whole numbers. For example, there is no number representing the difference $3 - 5$. New numbers called *negative numbers* are therefore introduced to make such operations possible.

These numbers are perhaps best appreciated when used to represent points on a scale or line situated below an arbitrary reference point or origin, such as 20 degrees below zero, 20 feet below sea level, or as —20 degrees, —20 feet, etc. In contrast, those above the reference point are then assumed positive and the set of both positive and negative numbers are then known as *directed numbers.* Here it is important to note that the signs '+' and '−' are used to distinguish the quality (or direction) of a number as well as to denote the operations of addition and subtraction on them. The sign of quality of a positive number (+) is usually omitted as 'understood' but the so-called 'rule of signs' ('like signs give plus, unlike give minus') becomes much easier to follow when referred to directed numbers in which every number concerned in an operation is enclosed in a bracket together with its sign of quality. In this way it is always possible to distinguish the sign of operation from the sign of quality.

For example, $(-2) + (-6) = (-8)$, and $(-2) + (+6) = (+4)$, and so on.

If the numbers are replaced by ratios, and expressed in logarithmic units, then the reference ratio will be $\frac{1}{1}$, or 0 cents, negative values will represent ratios of less than $\frac{1}{1}$, positive values ratios of more than $\frac{1}{1}$. Thus from the Ratio Table 20 cents = ratio 1·01162, while −20 cents = ratio ·98851 or $\frac{1}{1\cdot01162}$.

Moving on again—the result of the operation of evolution (root extraction) on a positive whole number cannot, in all cases, be expressed as a positive whole number, or as a quotient of two such positive whole numbers, so that *irrational numbers* such as $\sqrt{2}$, $\sqrt{3}$, etc., have been introduced. But not all irrational numbers involve roots, *e.g.*, $\pi = 3\cdot1415\ldots$, $e = 2\cdot718\ldots$ etc.

Rational approximations can be found to any real roots that are irrational, *e.g.* $\sqrt{2} = 1\cdot414\ldots$, $\sqrt{3} = 1\cdot732\ldots$.

The set of numbers obtained by adding the negative rational and irrational numbers to the number system of arithmetic (or the natural numbers) is called the *real system of numbers.* This is done in order to distinguish them from imaginary and complex numbers—which are not used in this book.

Graphically, any real number may be considered as locating a point to one side or the other of a fixed point 'o', all of which points lie on a straight line, assumed, if necessary, to be extendible without limit in either direction.

APPENDIX I

SPACE-TIME PATTERN THEORY OF HEARING

Summarizing, then, the pitch of a tone is determined both by the position of its maximum stimulation on the basilar membrane and also by the time pattern sent to the brain. The former is probably more important for the high tones and the latter for the low tones. The loudness is dependent upon the number of nerve impulses per second reaching the brain and possibly somewhat upon the extent of the stimulated patch. The experience called by psychologists 'volume' or 'extension' is probably identified with the length of the stimulated patch on the basilar membrane. This extension is carried to the brain and forms a portion of excited brain matter of a definite size. It is, then, this size that determines our sensation of the volume of a tone. The low pitched or complex tones have a larger volume, while the very high pitched tones have a small one.

The time pattern in the air is converted into a space pattern on the basilar membrane. The nerve endings are excited in such a way that this space pattern is transferred to the brain and produces two similar space patterns in the brain, one on the left and the other on the right side. Enough of the time pattern in the air is sent to each of these stimulated patches to make time of maximum stimulation in each patch detectable. So when listening to a sound with both ears, there are four space patterns produced in the brain, each carrying also some form of time pattern. It is recognition of the changes in these patterns that accounts for all the phenomena of audition.

(Taken from pp. 276 and 277 of *Speech and Hearing in Communication* by Harvey Fletcher, published in 1953, and quoted here by courtesy of the D. van Nostrand Company, Inc., of Princeton, New Jersey.)

Appendix 2

LISTENING CONTRAPUNTALLY

We are told that it is psychologically impossible to attend to many things at once. In fact, the psychologists tell us that it is a scientific fact that nobody can attend to more than two. This is quite true, but it has no effect whatever in limiting the complexities admissible in polyphonic music. No psychologist denies that you cannot hear any number of things at once. He denies only that you attend to more than two; but for my own part I seriously doubt whether in listening to polyphonic music I attend to more than one thing, and that is the total impression. From moment to moment I may notice an effect produced by this or that inner or outer part, but my enjoyment would be a sadly strenuous and uncomfortable affair if I made any conscientious effort to identify in a single hearing all the details which prove that the harmony and form are alive.

. . . . Triple counterpoint is harmony made by a combination of melodies so contrived that any of them can be a bass to the others. The necessity for its existence is evident in an art-form like that of a fugue; for, if any of the melodies is not capable of making a good bass to the others, the bass will be deprived of its fair share of the themes. The question whether the listener can attend to all three at once does not arise. He can recognise the whole combination, and he can enjoy the unity in variety which results from presenting the same harmonic and melodic elements in six different positions. As a matter of fact in Bach's standard type of triple counterpoint the three themes are remarkably transparent to each other. . . . There is usually one theme that skips energetically, another that trickles smoothly, and a third which completes and enriches the whole harmony with a slow chain of suspensions. . . .

(*The Integrity of Music* by Donald Francis Tovey, published by the Oxford University Press (1941), pp. 35-37.)

Few things demand more concentrated attention than accurate listening to a complex contrapuntal phrase. No one can follow several melodic lines simultaneously with anything like the completeness that ear and brain permit for a single melody. Yet musicians, not quite sure of physiological limitations, fear that the perfection of their hearing might be questioned, and make astonishing claims to super-acute powers. When we come to think of the mental qualities that would have to go with such ears, we simply find ourselves in dreamland. In any effort to listen contrapuntally, one has to expend most of one's energies in an admittedly exhausting process of rapid elimination and substitution. The total of one's impressions, in fact, depends on the alertness with which one can sustain this constantly changing accommodation and ceaseless readjustment of focus.

(*Down Among the Dead Men* by Bernard van Dieren, published by the Oxford University Press (1935), p. 189.)

Often a note in one of the parts (usually the bass) has to do a

Ex. 74. Ibidem, (Christe Eleison).

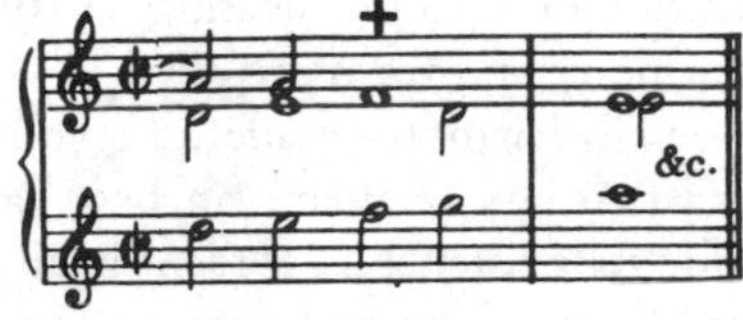

Ex. 75. Byrd, "I thought that Love".

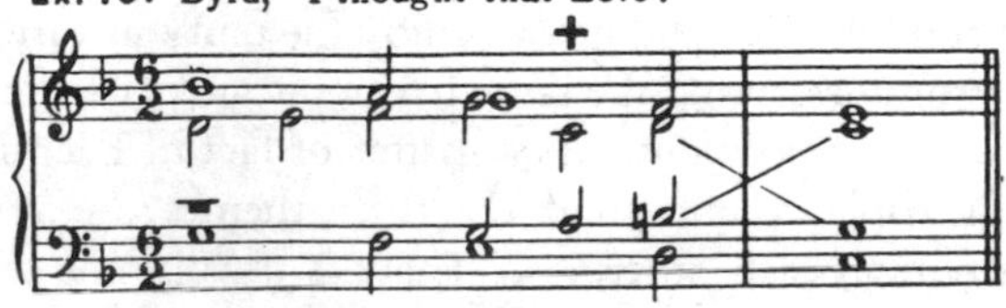

Ex. 76. Idem, "Have Mercy upon me".

kind of double duty. Take the progression shown by Ex. 74 (the passage quoted actually occurs in *Assumpta est Maria*, but this and similar progressions are exceedingly common in all the writings of the sixteenth century). It is impossible to explain this chordally in terms that would have been intelligible to Palestrina; but the real explanation is very simple. The B in the bass is a passing note from the point of view of the soprano, but as far as the tenor is concerned it is a harmony note. Similarly the E in the tenor is a discord against the F of the soprano (and its resolution is normal), but tenor and bass are in concord on each beat. The whole passage must be regarded as a complex of three two-part progressions, each of which is in itself perfectly normal and regular, and which may, therefore, be employed in combination. Exx. 75 and 76 afford further illustrations of this in four and five parts. In his own writing, whenever doubtful as to the rightness of a particular progression, the student should apply this test of horizontal analysis, and if the progression stands the test, he can use it without fear. It is obvious that by working on such principles he will obtain (as the sixteenth-century composers obtained) a considerably wider range of harmonic freedom than the text-book rules allow.

(*Contrapuntal Technique in the Sixteenth Century*, by R. O. Morris, p. 37—first published in 1922, and reproduced here by courtesy of The Clarendon Press at Oxford.)

An important element in imitation is the amount of time which elapses between the successive entries of the voices. It is rather rare in the sacred music of this time to find entries spaced regularly—that is, to find successive voices entering regularly one or two measures apart. This is flat rhythmically, and the sixteenth century rarely allowed such mechanical devices to recur. In music as in architecture artists were careful to vary their designs so that successive ideas would not appear machine-made.

If a theme in the beginning voice enters on the first beat of the measure (as it usually does), it ordinarily does not begin on the first beat in the next voice, either in the second measure or the third, but in the middle of one of these measures, as in Ex. 158.

In this example there is a wonderful plasticity of rhythm in the way the successive voices enter: the alto on the first beat of

1
3
6
9
11
15
etc.

measure 1, the soprano on the third beat of measure 3, the tenor on the first beat of measure 6, the bass on the first beat of measure 9, the soprano on the third beat of measure 11, and the alto on the first beat of measure 15. Such irregularity in entries aids tremendously in keeping the rhythm of the composition plastic, as well as in making each voice individual and not simply a cog in a wheel.

An example of a short piece built on only one theme is the Kyrie I of Palestrina's Mass *Aeterna Christi munera* (Ex. 159). The piece is only thirteen measures long. The tenor enters on the

first beat and the alto on the third beat of the first measure; the first note of the soprano is halved so that it can enter in the second measure (it could not enter on the third beat on account of the harmony there, and therefore enters on the fourth beat); the bass waits until measure 6 to come in; the soprano enters again at the beginning of measure 8; the alto begins again on the fourth beat of the same measure with a half note; and the tenor enters a second time on the fourth beat exactly a measure later.

Let us make one more observation in regard to the entries of voices in imitation. Under ordinary circumstances an entering voice is more effective if it makes its entry in a register which has been unused by other voices for a time. This is well illustrated by the example, *Jesus junxit*. The first entry of the soprano stands out because the D has not been heard before; the tenor entry stands out because this D has not been heard for over two measures; the bass initial has never been heard up to this point; the next soprano entry is clear, for no voice has sung this note since the soprano herself left off with it a measure and a half before; and the next alto entry is on an A which was only touched a measure before by the soprano. This type of writing is always effective in giving clarity to the individual voices, and for the beginner it offers a valuable principle to follow.

In this composition the way the lines behave in regard to each other is worth noticing, because it is so typical of the period. For instance, the crossing back and forth between the soprano and the alto in measures 21 and 22 and in measures 26 and 27 not only results in long lines which flow smoothly but also gives a remarkable play of voice color. The tone quality of the soprano and alto voices is different by nature, and in such passages as those just cited the two voices have particular charm and beauty, for neither loses its individuality in the process of weaving; now the lighter tone quality is on top, now underneath, now again on top. Played on the piano, such a passage is merely stupid, since there is no difference in tone color, but with two orchestral instruments of different tone colors the same differentiation becomes apparent as with voices. The effect is magnificently exploited by Bach in such pieces as the trio-sonatas for organ, where each line must be played on a separate keyboard. While this device of crossing voices for the sake of tone color is by no

20
glo - ri - a tu -
- a tu - - - a, glo - ri - a tu - -
tu - a,
25
- a, glo - ri - a
- - a, glo - ri - a tu - - -
glo - ri - a tu - a,
30
tu - a glo - ri - a tu - a, glo -
- - - - - a, glo - ri - a
glo - ri - a tu - a,

means rare in sacred music of the sixteenth century, it is in secular music of the period that it is used most often; there it really flowers.

(The above have been extracted from pp. 101 to 104, and from pp. 159 and 162 of Arthur Tillman Merritt's book entitled *Sixteenth-Century Polyphony*, and is reproduced here with the generous consent of the Harvard University Press, who published it in 1964.)

APPENDIX 3

CHORAL MUSIC AND CHROMATICS

The 'theoretician's' faith in equal temperament as a musical solvent has often misled others. It has no basis in reality. Unaccompanied voices encounter an unfair task when they attempt to sing the production of some uninstructed writer with a keyboard mind, who has sprinkled his score with chromatics under the impression that there is no difference between F♯ and G♭, and that all semitones, chromatic or diatonic, are equal. The result is almost sure to be an uncertain intonation and an uncomfortable loss of pitch. The judicious use of occasional chromatics may present no great difficulty to a good choir, but the musician has much to learn before he can write gratefully for voices. Even the scholarly Hauptmann (1792-1868), on occasion, set his choir a task which at the time of writing he failed to realize was beyond them. A piece of self-criticism, to be found in a letter which he wrote in 1843 to Spohr (who made free use of chromatics), shortly after he became Cantor of the *Thomas-schule* in Leipzig, is pertinent to any discussion of the musical scale:

'I hope we shall soon venture on your Psalms; I was afraid to begin with them. The Chorus [of the *Thomas-schule*] is firm as a rock in diatonics, with figures and *colorature ad libitum*, but in chromatic music, they are no better than their fellows. To sing chromatic passages in tune, presupposes a real education in music; something more than hitting the note is required; the singer must feel for himself the harmonic progressions. This I learn, to my mortification, every time that I hear this passage in my *Salve Regina* [an early work, Op. 13]:

Mechanically, there seems to be nothing amiss, but when it comes to the performance, I am always in Purgatory. The sharpness does not lie in the vocal intonation. . . . There is no justification for a composer who makes a pianoforte accompaniment indispensable for the performance of choral music; and the old masters were far from wrong, in adhering to a very peremptory code of laws, to regulate such compositions as this. I am more ashamed of such a passage, than I should be of palpable octaves and fifths, which anyhow are no hindrance to pure intonation.'

(Reprinted here from Ll. S. Lloyd's *The Musical Ear*, by courtesy of the publishers, Oxford University Press.)

APPENDIX 4

VARIOUS OBSERVATIONS CONCERNING INTONATION AND INSTRUMENTS

The Oboe

'Fine intonation is vital on any instrument, but since the natural quality of the oboe tone is so concentrated and clearly defined, imperfect intonation on the oboe sounds particularly painful, and you must guard against it with all possible care. Playing well in tune depends on the ear's ability to judge the correct intonation of each note, and on the control of the muscles necessary to adjust the pitch according to the ear's dictates. Some players are born with a better musical ear than others, but any musical ear—even a poor one—can be improved by concentrated practice. Always listen to yourself with the most critical attention, for playing out of tune can so easily become a careless habit which spoils the performance even of accomplished players. Vibrato can camouflage this fault, so practise slow scales and long notes *without* it. Try always to hear the note you are going to play before you actually play it: this should very soon become automatic, even when sight-reading.

As I have explained, the control of intonation and dynamics is achieved by *slight* alterations to the volume and pressure of the air stream, and these alterations are managed by the *co-ordinated* actions of the breathing muscles and the embouchure muscles. Always bear in mind this fundamental principle. . .

There are few oboes in existence—if any!—which do not have at least one or two slightly faulty notes, and these will have to be 'humoured' by the embouchure to be perfectly in tune. But *all* embouchure adjustments are so very sensitive and small that if you listen carefully to your intonation, your muscles will soon learn to adjust the embouchure automatically without any conscious thought on your part.

(Quoted, by kind permission of the author, from pp. 24 and 25 of the second edition of *Oboe Technique* by Evelyn Rothwell, and published by Oxford University Press in 1962.)

APPENDIX 4

The Cello

'Intonation,' Casals told a student, 'is a question of conscience. You hear when a note is false the same way you feel when you do something wrong in life. We must not continue to do the wrong thing.' His assertion that 'each note is like a link in a chain—important in itself and also as a connection between what has been and what will be', applied as equally to intonation as to other aspects of interpretation. The notes of a composition do not exist in isolation; the movement of harmonic progressions, melodic contours and expressive colorations provides each interval with a specific sense of *belonging* and/or *direction*. Consequently, Casals stressed that the equal-tempered scale with its fixed and equidistant semitones—as found on the piano—is a compromise with which string players need not comply.[1] Playing in tune is therefore not a matter of adherence to intervals based upon a pre-ordained mathematical formula; it is a dynamic process, expressing the organic relationship between notes in a musical context, which Casals termed 'expressive intonation'. The final judgment lies in the ever-sensitive ear of the musician.

Because it is a natural and instinctive response to music, expressive intonation is to some extent practised spontaneously by many musicians. However, few apply it with the comprehensive awareness that characterized Casals' approach. New students coming to Casals—most of them advanced, some already professional—would often have comfortable illusions shattered when their habitual intonation was challenged by his uncompromising ear.

The precise intonation of semitones will also be affected by the speed at which they are performed. Casals advised, 'In a relatively fast movement [i.e. when the specific semitone relationships move quickly] we have to exaggerate still more the closeness of the half-tones.'

When demonstrating a properly measured semitone, Casals would sometimes exclaim, 'Isn't it beautiful!' And indeed it was. The placement of intervals in meaningful relativity provides a

[1] These remarks are also applicable to wind players and, not least, to singers.

fundamental sense of well-being. The notes fall into place with inevitability, thus gaining in vitality. Intellectual awareness, intuitive perception and critical listening all play a role in the determination of the precise degree to which the instrumentalist adjusts his pitch.

Expressive intonation, when observed continuously throughout a composition, becomes a foremost factor in the communication of emotional content.

Casals considered it essential that expressive intonation be taught to string players from the beginning of their studies. He took endless trouble in retraining the aural sense and habitual finger placements of students who, since childhood, had unquestioningly applied piano intonation to their stringed instruments. 'The effects of any neglect of this kind at the beginning of studies . . . can affect a player through the whole of his career, however gifted he may be.' I once met the living proof of this statement in a cellist who was attending Casals' Berkeley classes—a performer not without talent but who had early on been brainwashed by equal temperament. Hearing Casals for the first time, she exclaimed, 'It is *soooooo* beautiful—but why does he play out of tune?'

The degree of bow pressure and the volume at which one plays are relevant to intonation. Where intensive bow pressure is indicated, the string will tend to sharpen and a necessary compensation must be made by the left hand.

(Taken from pp. 102, 103 and 106 to 109 of *Casals and the Art of Interpretation* by David Blum, published in 1977, and quoted here by kind permission of Holmes and Meier of New York, and Heinemann of London.)

The Violin

Lastly, in this discussion of intonation, it is necessary to consider what type of intonation ought to be used: the 'tempered' or the 'natural'. This is not the place to go into the technicalities of the two systems. No violinist can play according to a mathematical formula; he can only follow the judgment of his own ear. Be this as it may, *no one system of intonation will suffice alone.* A performer has constantly to adjust his intonation to match his accompanying medium.

The artist must be extremely sensitive and should have the ability to make instantaneous adjustments in his intonation. (The best and easiest way to make such adjustments is by means of the vibrato.) An intonation adjustable to the needs of the moment is the only safe answer to the big question of playing in tune.

The most important part in all of this is assigned, obviously, to the ear, which has to catch immediately the slightest discrepancy between the pitch desired and the pitch produced and then demand an instant reaction from the fingers.

Advanced players, already in possession of a secure intonation, will find that their facility for quick adjustment can be improved further by changing from time to time the instruments they use. It is also good advice not to interrupt the practise every few minutes to retune the violin. One should be able to play *in tune* on a violin which is *out of tune*. The performer who has acquired such a skill will never be shaken out of his assurance and authority in public performance by a recalcitrant string.

(Taken from p. 22 of the *Principles of Violin Playing and Teaching* by Ivan Galamian—published in 1962, and reproduced here by courtesy of the publishers, Prentice-Hall of Englewood Cliffs, N.J., U.S.A.)

The Trombone

Musically, the free-slide principle has endowed the trombone with a refinement which is found nowhere else in the orchestra save among the unfretted strings. As every position of the extended slide is finally determined by ear (quite instinctively by the experienced player) extremely subtle degrees of inflection are possible, and the trombonist can temper the intervals between notes to an extent beyond the reach of other wind-players. Indeed, it has been said that the trombone is the only wind instrument that can be played completely in tune all the time. The converse is, of course, unfortunately also true.

(Quoted—by kind permission of the author—from p. 51 of the second edition of *The Trumpet and Trombone*, by Philip Bate, and published by Ernest Benn of London and W. W. Norton of New York.)

The Horn

However good the design of an instrument and however accurate its manufacture, in the last resort perfect intonation is to a large extent in the hand of the player. A well-designed and carefully made horn will make it easier for the player to secure correct intonation, but all this will be of little avail unless he has an ear capable of distinguishing between accurate and inaccurate intonation, and the ability to correct small imperfections by positive action on his own part. Too often it is assumed that all the player of a wind instrument has to do is to press the right keys or valves and an accurately tuned note automatically follows. This is very far from being the case, but with the horn two methods are open to the player of correcting his intonation. One, the manipulation of the right hand in the bell, has already been described, but is rarely necessary on the wide-bore instrument. The other, possible only when a small correction is required, since it will otherwise have an adverse effect on the tone quality, is to force the air column, by slight contraction or relaxation of the lip muscles, to vibrate at a frequency slightly different from that determined by its dimensions. Such a process, known as 'lipping', is employed by expert players to bring into correct intonation those notes in the instrument which are inherently slightly out of tune, and the adjustment eventually becomes automatic. Inadvertent lipping where it is not required is more then likely in the case of the inexpert player with a faulty embouchure and inadequate muscular control.[1]

[1] A recent book puts these points somewhat infelicitously as follows: 'It is possible to force the natural resonance a little one way or the other by the lips, and the pitch range can be slightly altered by the insertion in the mouth of a large pear-shaped mute, or even the hand.'

Finally, it appears to be a fact verified by experiment that players of wind instruments in which small modifications of intonation are possible do not actually keep to any one of the standard scientific systems of intervals, any more than players of stringed instruments do. Though in theory the pitch of any particular note is absolute, in practice this is not so, for much depends upon its context. Playing in tune, then, is not so much a question of adhering to a fixed system as of careful adjustment to the sounds of other instruments. In the last resort, what sounds right is of greater importance than the exact frequency of the note sounded.

(The above has been extracted from p. 73 of the second edition of Robin Gregory's book entitled *The Horn*, and is reproduced here with the generous consent of Messrs. Faber and Faber, who published it in 1969.)

The Old Wood-wind Instruments

The intonation of the recorder right through the chromatic compass of two octaves and one note is perfect, if you know how to manage the instrument; but its fingering is complicated, and requires study. To the ignorant person who just blows into it, and lifts one finger after another to try the scale, it seems horribly out of tune; but that is not the fault of the instrument.

This brings about the whole question of the alleged imperfections of the old wood-wind instruments. The 18th century one-keyed flute, for example, has been tried more than once in recent years, and in every case the same verdict was returned: that its tone, whilst it is inferior in power to the modern flute, especially in the low notes, is much more beautiful and characteristic, but that its intonation is defective.

This last sentence is not true. The one-keyed flute has been thoroughly and patiently studied recently by a flautist who followed the instructions contained in the 'Principes de la Flûte Traversière,' by the famous Hotteterre le Romain, Ordinaire de la Musique du Roy, Paris, 1707, with the result that he can now play on the old flute more perfectly in tune than he ever did before upon a highly improved and most expensive modern instrument. And the reason is not far to seek. On the old flute, almost every note has to be qualified by the breath, or some trick of fingering, or the turning of the flute inward or outward, to cover more or less of the *embouchure*, by which means the pitch of the notes can be affected to a great extent. This requires the constant watchfulness of the ear, which thus becomes more and more sensitive to faults of intonation.

On the modern flute, a most ingenious and complicated system of keys has been devised, which is supposed, at the expense of beauty of tone, to correct automatically all the imperfections. Under ideal conditions, it might come near doing this. But in practice the instrument may be flat from being cold, or sharp from being warm, or be affected by the variations of wind-pressure necessary for the lights and shades, and the player whose ear is not only untrained but hardened by the instruction he has received, goes on playing out of tune, often to a most painful extent, without feeling in the least distressed, for the simple reason that trusting to his instrument, he does not even listen to what is going on around him, which is the absolute condition to obtain such effects as can satisfy a discriminating auditor.

There are in the old books innumerable instructions, rules and warnings, intended to foster pure intonation in all instruments. Quantz's 'Versuch' is full of such; everything concurs to prove that the old musicians were extremely sensitive on that point.

(Copied from pp. 357 to 459 of *The Interpretation of the Music of the XVII and XVIII Centuries Revealed by Contemporary Evidence*, by Arnold Dolmetsch—first issued in 1916—published, in 1946, and reproduced here by kind permission of Messrs. Novello & Company Limited of Sevenoaks, Kent.)

The Clavichord

Beethoven said: 'Among all keyed instruments the clavichord was that on which one could best control tone and expressive interpretation.'[1] In his day the clavichord was going out. But it was a favourite instrument of J. S. Bach, and is particularly

[1] Cited in Grove's *Dictionary of Music and Musicians* (1927 ed.), i, 662.

well-suited to many of his immortal Forty-eight Preludes and Fugues.

The clavichord did not come into prominence before the sixteenth century, though the principle is much older, and the instrument itself may date before the fourteenth century. This principle is what is called tangent action.

The tangent is a brass wedge set upright at the far end of a pivoted wooden lever. When you depress the near end (the visible key), the tangent is raised smartly into contact with the string (or often course of strings: for a unison pair is usual, the material being covered or uncovered brass).

The string is divided into unequal lengths, of which the shorter is kept voiceless by interwoven strips of felt. The longer, however, is caused by the impact to vibrate, giving a note predetermined by the weight, length and tension of the string in question.

But not wholly predetermined. For though the tension has been set within very fine limits by adjusting the tuning-pins in the ordinary way, within still finer limits the player can control the tension directly, by pressing more or less firmly on the key.

This gives him something of a violinist's direct finger control over that expressive, faintly but regularly fluctuating variation of intonation which is called vibrato. But when he releases the key the sound is instantaneously damped.

Now in apparent theory, since every increase in pressure must needs raise the pitch, it ought to be impossible to vary the expression without playing out of tune. This has misled some historians (perhaps prejudiced already by hearing a bad clavichordist) into deducing that the clavichord is inherently defective. But it is the theory which is defective, not the instrument.

By striking more sharply, you can heighten the *velocity* of the tangent, thus increasing the loudness, without appreciably altering the pitch. If, however, you are clumsy, you may then exaggerate the *pressure*, thus playing out of tune. This is especially liable to happen to pianists casually essaying the clavichord. For the touch of these instruments is radically different: and it is impossible for the pianist to unlearn in a moment the almost unconscious muscular habits acquired by years of drawing the best results from his own instrument. Even if he does not play the clavichord out of tune, he cannot hope to make it speak in its finest tones without proper study.

When well played, the clavichord, though less brilliant than the harpsichord and less massive than the piano, is, as Beethoven

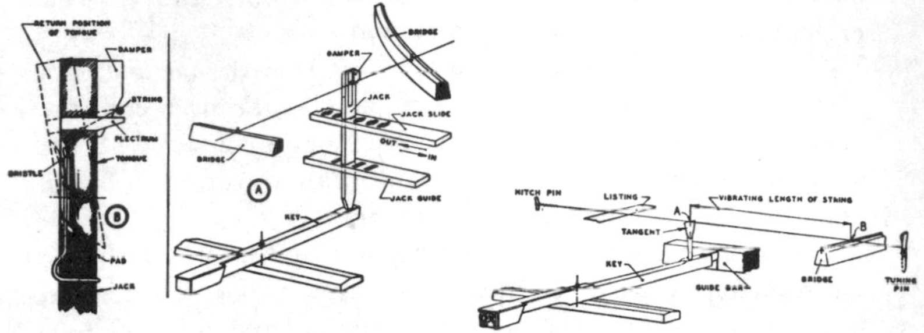

noted, more subtly expressive than either. Its loudest tones are very soft, but its softest tones are unbelievably softer still; and your control over this whole dynamic range is, from the nature of the action, more intimate than any other keyboard instrument affords. And, indeed, a (relatively) loud chord on the clavichord sounds more explosive than the (absolutely) loudest chord of the mighty piano itself, so that those sudden outbursts, for example, which punctuate the closing passages of J. S. Bach's Chromatic Fantasia acquire, by sheer contrast, a tragic power and horror on the tiny clavichord which no concert grand piano can approach.

Again in fugal music, to make a fugal entry stand out clearly on the piano, you have to play it decidedly more loudly than the accompanying parts. But on the clavichord the tone is of such clarity that the counterpoint is more easily made distinct, while the fugal entries can be given a more expressive nuance by direct finger control, as on no other keyboard instrument. And finally, the sensuous beauty of good clavichord tone is itself very moving. But because of its overall softness, the clavichord is essentially unsuitable for public performances; nor is it satisfactory when electronically amplified. It can be recorded well—if played with the volume low.

(Reprinted here—by courtesy of the author—from pp 96 to 98 of the third edition of *The Instruments of Music*, by Robert Donington, and published in 1970 by Methuen & Co. Ltd., of London.)

Appendix 5

THE 'CHORD OF NATURE'

An extract from Ll.S. Lloyd's article on 'Modern Science and Musical Theory', reprinted with kind permission from the *Journal of the Royal Society of Arts*, Vol. XC, No. 4619, of August 7th, 1942.

On the left of Fig. 5 is the so-called chord of nature, consisting of the 4th, 5th, and 6th harmonics of C below the bass clef. These harmonics are, of course, pure tones. The use of the word 'chord' suggests to unsuspecting musicians that it is identifiable with the chord of (white) musical notes I have shown on the right. But, played on musical instruments, say three violins, each of the notes in this chord contains its own chain of overtones. Let us set out these white notes separately, as in the figure, where I have shown a complete series of six harmonics (the prime and five overtones) for each one; for on strings there would be sensible vibrations in the air corresponding to each harmonic.

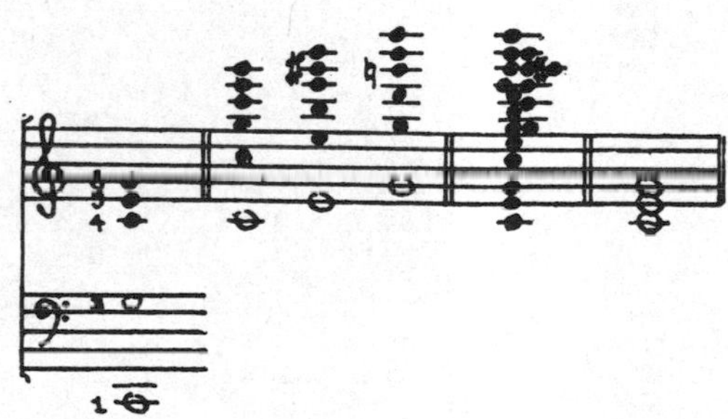

Fig 5.

Now let us combine all these notes. I do not find it very convincing to be told that the group of black notes on the left is a chord of nature when actually 'nature' sets in motion all the vibrations in the mass of black notes on the right when artists want the musician to hear a chord.

The really interesting thing is the reaction of the ear in the two cases. We known that the ear extracts the three notes shown in semibreves from the mass on the right. I don't think that anyone knows, quite completely, how it is done. But now take the 'chord of nature'. Fletcher has made many experiments, using pure tones produced electrically in the laboratory, with three or more consecutive harmonics, the prime being absent. His results tell us that with three harmonics we should probably hear the bass C. If we make four successive harmonics by adding the third harmonic, which I have dotted in, Fletcher's results tell us that we should certainly hear the bass C.

As Helmholtz himself observed, 'the system of scales, modes and harmonic tissues does not rest solely upon unalterable natural laws, but is also, at least partly, the result of aesthetical principles, which have already changed, and will still further change, with the progressive development of humanity'.

APPENDIX 6

THOROUGH-BASS, OR THROUGH-BASS, AND THE THEORY OF THE FUNDAMENTAL BASS

The great classical tradition cares little for the study of chords as things in themselves; and the art of harmony perishes under a discipline that separates its details from counterpoint and its larger issues from form. An excellent means of mastering a good harmonic vocabulary is to practise the filling-out of classical figured basses at the keyboard; in other words, to exercise the function of the continuo-player who, from the time of Monteverdi to that of Beethoven's organ-teachers, used to supply accompaniments from a bass with figures indicating the gist of the chords required. Fluency in such a practice does not of itself confer the ability to produce original harmony, but it means that music can be read with understanding. It is an empiric craft. But it had the misfortune to become a science, when, early in the eighteenth century, Rameau discovered the theory of the fundamental bass. This is an imaginary bass (best when most imaginary) that gives 'roots' to all the essential chords of the music above it. The conception is true only of the most obvious harmonic facts; beyond them it is as vain as the attempt to ascertain your neighbour's dinner from a spectrograph of the smoke from his chimney. The augmented sixth which arose so innocently in Ex. 12 requires a double root. The first chord of

Beethoven's Sonata in E flat, op. 31, no. 3, is an 'eleventh' with its root on the dominant in flat defiance of the fact that the dominant is the most inconceivable bass-note in the whole

passage until it arrives as a climax in the sixth bar. But musical fundamentalists refuse to look six bars ahead.

(Taken from pp. 61 to 63 of *Musical Articles from the Encyclopaedia Britannica* by Donald Francis Tovey, published in 1944, and quoted here by kind permission of the Oxford University Press.)

APPENDIX 7

THE TERM 'WOLF' (as used in tuning)

Some very good tuners will help a little, by robbing Peter to pay Paul; as by making ♯G over sharp. But then E, a more usual note, will suffer in its ♯3rd, which will hurt the musick in *A* key; and for that reason they call that note [G♯] the wolf, which may neither be held, nor let go[15].

[15]The phrase 'to have a wolf by the ears' (so that one is both unable to. hold it and afraid to let it go) occurs in ancient Greek authors and in many later ones. Among musicians, Schlick in 1511 mentioned 'the discordances which the organ-maker calls wolves', and most writers have assumed that the word was adopted because mistuned intervals might (with some imagination) be said to sound like wolves. North's remark suggests an alternative usage, in which G♯ is a 'wolf' because it puts the tuner in a dilemma, not because of any supposed 'howling' of the false fifth G♯—E♯. He also refers to other notes 'upon which the scismes clutter most, as E♭ and ♯C, which will be wolvish, unless more usefull accords are wounded'.

(Extracted, complete with footnote, from p. 211 of *Roger North on Music*, transcribed and edited by John Wilson and published, in 1959, by Novello and Company Ltd—by whose courtesy it appears here.)

Appendix 8

DIFFERENCE TONES FROM THE HARMONIC SERIES

The ratios of the frequencies of the partial vibrations of a perfectly uniform flexible string, or of any maintained vibration, fall into the series

1 2 3 4 5 6 etc.

so that the corresponding musical intervals between, say, the 1st and 2nd of this series, ratio $\frac{2}{1}$, is an octave—that between the 2nd and 3rd, ratio $\frac{3}{2}$, a fifth—and so on.

This series provides, among other things, a useful means by which the position of the difference tone for any given interval may be easily obtained from its ratio.

For instance, the difference tone for—

the octave, ratio $\frac{2}{1}$, corresponds to $2 - 1 = 1$
i.e., makes a unison with the lower note (*i.e.* '1').

the fifth, ratio $\frac{3}{2}$, corresponds to $3 - 2 = 1$
i.e., is an octave below the lower note (*i.e.* '2').

the fourth, ratio $\frac{4}{3}$, corresponds to $4 - 3 = 1$
i.e., is a twelfth below the lower note (*i.e.* '3').

the major third, ratio $\frac{5}{4}$, corresponds to $5 - 4 = 1$
i.e., is a double octave below the lower note (*i.e.* '4').

the minor third, ratio $\frac{6}{5}$, corresponds to $6 - 5 = 1$
i.e., is a double octave plus a major third below the lower note (*i.e.* '5').

the major sixth, ratio $\frac{5}{3}$, corresponds to $5 - 3 = 2$
i.e., is a fifth below the lower note (*i.e.* '3').

the minor sixth, ratio $\frac{8}{5}$, corresponds to $8 - 5 = 3$
i.e., is a major sixth below the lower note (*i.e.* '5').

etc.

All the above results may be verified visually by observing the positions of the intervals between the partials of the harmonic series as indicated by their spacings on any single one of the slide-rules shown in Fig. A of Chapter 12.

This method can also be used to find the positions of the difference tones of any triad (or chord) by considering its component intervals, one at a time. See also pp. 18 and 19.

APPENDIX 9

THE USE OF CONTINUED FRACTIONS

Decimal ratios or ratios involving large numbers are sometimes required to be expressed, more or less accurately, in terms of the ratio of two whole numbers of a reasonably low value. The reasons why such ratios are needed include—the critical study of different temperaments, particularly the cyclic types—the production of such tunings on electro-mechanical devices employing chains of mechanical gears—the theoretical positioning of frets by geometrical means—etc.

In general the greatest possible accuracy is sought from the lowest possible ratio. The method of continued fractions—described below—is one way of meeting this demand, for, not only does it give the required ratios in increasing order of accuracy, starting from the lowest, but also in terms of the lowest possible numbers for a given accuracy.

In order to demonstrate this method, the decimal ratio of an equally-tempered semitone, correct to five significant figures, *i.e.*, 1·0595, or more correctly, $\frac{1\cdot0595}{1}$, has been chosen. The problem is to find ratios of numbers lower than 10595 and 10000 whose value approximates to this ratio.

$$\frac{1\cdot0595}{1} = \frac{1}{1} + \frac{595}{10000} = 1 + \frac{119}{2000} \quad \ldots \quad \text{(a)}$$

$$\text{but } \frac{119}{2000} = \frac{1}{\frac{2000}{119}} = \frac{1}{16+\frac{96}{119}} \quad \ldots \quad \ldots \quad \ldots \quad \text{(b)}$$

$$\text{and } \frac{96}{119} = \frac{1}{\frac{119}{96}} = \frac{1}{1+\frac{23}{96}} \quad \ldots \quad \ldots \quad \ldots \quad \ldots \quad \text{(c)}$$

$$\text{and } \frac{23}{96} = \frac{1}{\frac{96}{23}} = \frac{1}{4+\frac{4}{23}} \quad \ldots \quad \ldots \quad \ldots \quad \ldots \quad \text{(d)}$$

$$\text{and } \frac{4}{23} = \frac{1}{\frac{23}{4}} = \frac{1}{5+\frac{3}{4}} \quad \ldots \quad \ldots \quad \ldots \quad \ldots \quad \text{(e)}$$

and so on.

This means that

$$\text{from (a) } \frac{1\cdot0595}{1} = 1 + \frac{119}{2000}$$

$$\text{from (b) } \quad = 1 + \frac{1}{16+\frac{96}{119}} \quad \ldots \quad \ldots \quad \ldots \quad \text{(f)}$$

$$\text{from (c) } \quad = 1 + \cfrac{1}{16+\cfrac{1}{1+\frac{23}{96}}} \quad \ldots \quad \ldots \quad \ldots \quad \text{(g)}$$

from (d) $= 1 + \cfrac{1}{16 + \cfrac{1}{1 + \cfrac{4}{4 + \cfrac{4}{23}}}}$ .. .. .. (h)

from (e) $= 1 + \cfrac{1}{16 + \cfrac{1}{1 + \cfrac{1}{4 + \cfrac{1}{5 + \frac{3}{4}}}}}$ .. .. .. (i)

and so on.

By neglecting the last fraction to appear on the right the following successive approximations may be obtained:

from (f) $1 + \frac{1}{16} = \frac{17}{16}$ (1·0625)
from (g) $1 + \frac{1}{17} = \frac{18}{17}$ (1·0588)

from (h) $1 + \cfrac{1}{16 + \cfrac{1}{1 + \frac{1}{4}}} = 1 + \cfrac{1}{16 + \cfrac{1}{\frac{5}{4}}} = 1 + \cfrac{1}{16 + \frac{4}{5}} = 1 + \cfrac{1}{\frac{84}{5}} = 1 + \frac{5}{84}$

$= \frac{89}{84}$ (1·05952)

etc.

The decimal fractions quoted in brackets above give some idea of the degree of approximation in each case, but to get a true picture logarithmic units are necessary. Below is listed the series of approximations obtained above together with the amount of their deviations (in cents) from that of an exact equally-tempered semitone of 100 cents.

$\frac{17}{16}$ (+4·955) $\frac{18}{17}$ (−1·045) $\frac{89}{84}$ (+·099) etc. .. .. (j)

These do not include all the possible ratios, and since any temperament involves a good deal of compromise one way or another, many ratios other than those above have to be considered—though as already stated—such ratios must in general involve higher numbers for the same degree of accuracy.

The following principle enables such additional ratios to be readily derived from those listed at (j):—

If the numerators and denominators of any two fractions (or ratios) are each added together separately so as to form a new fraction, then the new fraction (or ratio) has a value which lies between those of the two fractions (or ratios).

For example

$\frac{3+4}{4+5} = \frac{7}{9}$ (= ·7777) which lies between $\frac{3}{4}$ (= ·75) and $\frac{4}{5}$ (= ·8)

The application of this principle to the series of approximations at (j) yields the following ratios:

$\frac{35}{33}$ (+1·87) $\frac{107}{101}$ (−·093) etc. .. .. (k)

The property of fractions stated in Chapter 19 is sometimes of value in comparing the relative magnitude of two approximately equal ratios.

Suggested Book List For Further Reading

FR1 DONINGTON, ROBERT. *The Instruments of Music*, 2nd ed., Methuen & Co. Ltd., London, 1970.

FR2 BENADE, ARTHUR H. *Fundamentals of Musical Acoustics*, 1st ed., Oxford University Press, London, 1976.

FR3 JORGENSEN, OWEN. *Tuning the Historical Temperaments by Ear*, 1st ed., The Northern University, Marquette, 1977.

FR4 TOVEY, DONALD FRANCIS. *Musical Articles from the Encyclopaedia Britannica*, 1st ed., Oxford University Press, London, 1944.

POWERS, ROOTS AND RECIPROCALS

$\frac{1}{1}$	1·0	$\frac{1}{26}$	·038462	$\frac{1}{51}$	·019608	$\frac{1}{76}$	·013158
$\frac{1}{2}$	·50000	$\frac{1}{27}$	·037037	$\frac{1}{52}$	·019231	$\frac{1}{77}$	·012987
$\frac{1}{3}$	·33333	$\frac{1}{28}$	·035714	$\frac{1}{53}$	·018868	$\frac{1}{78}$	·012821
$\frac{1}{4}$	·25000	$\frac{1}{29}$	·034483	$\frac{1}{54}$	·018519	$\frac{1}{79}$	·012658
$\frac{1}{5}$	·20000	$\frac{1}{30}$	·033333	$\frac{1}{55}$	·018182	$\frac{1}{80}$	·0125
$\frac{1}{6}$	·16667	$\frac{1}{31}$	·032258	$\frac{1}{56}$	·017825	$\frac{1}{81}$	·012346
$\frac{1}{7}$	·14286	$\frac{1}{32}$	·03125	$\frac{1}{57}$	·017544	$\frac{1}{82}$	·012195
$\frac{1}{8}$	·12500	$\frac{1}{33}$	·030303	$\frac{1}{58}$	·017241	$\frac{1}{83}$	·012048
$\frac{1}{9}$	·11111	$\frac{1}{34}$	·029412	$\frac{1}{59}$	·016949	$\frac{1}{84}$	·011905
$\frac{1}{10}$	·1	$\frac{1}{35}$	·028571	$\frac{1}{60}$	·016667	$\frac{1}{85}$	·011765
$\frac{1}{11}$	·090909	$\frac{1}{36}$	·027778	$\frac{1}{61}$	·016393	$\frac{1}{86}$	·011625
$\frac{1}{12}$	·083333	$\frac{1}{37}$	·027027	$\frac{1}{62}$	·016129	$\frac{1}{87}$	·011494
$\frac{1}{13}$	·076923	$\frac{1}{38}$	·026316	$\frac{1}{63}$	·015873	$\frac{1}{88}$	·011364
$\frac{1}{14}$	·071429	$\frac{1}{39}$	·025641	$\frac{1}{64}$	·015625	$\frac{1}{89}$	·011236
$\frac{1}{15}$	·066667	$\frac{1}{40}$	·025	$\frac{1}{65}$	·015385	$\frac{1}{90}$	·011111
$\frac{1}{16}$	·0625	$\frac{1}{41}$	·02439	$\frac{1}{66}$	·015152	$\frac{1}{91}$	·010989
$\frac{1}{17}$	·058825	$\frac{1}{42}$	·02381	$\frac{1}{67}$	·014925	$\frac{1}{92}$	·01087
$\frac{1}{18}$	·055556	$\frac{1}{43}$	·023256	$\frac{1}{68}$	·014706	$\frac{1}{93}$	·010753
$\frac{1}{19}$	·052632	$\frac{1}{44}$	·022727	$\frac{1}{69}$	·014493	$\frac{1}{94}$	·010638
$\frac{1}{20}$	·05	$\frac{1}{45}$	·022222	$\frac{1}{70}$	·014286	$\frac{1}{95}$	·010526
$\frac{1}{21}$	·047619	$\frac{1}{46}$	·021739	$\frac{1}{71}$	·014085	$\frac{1}{96}$	·010417
$\frac{1}{22}$	·045455	$\frac{1}{47}$	·021277	$\frac{1}{72}$	·013889	$\frac{1}{98}$	·010309
$\frac{1}{23}$	·043478	$\frac{1}{48}$	·020833	$\frac{1}{73}$	·013699	$\frac{1}{99}$	·010101
$\frac{1}{24}$	·041667	$\frac{1}{49}$	·020408	$\frac{1}{74}$	·013512	$\frac{1}{100}$	·01
$\frac{1}{25}$	·04	$\frac{1}{50}$	·02	$\frac{1}{75}$	·013333		

$2^2 = 4^1 = 4$	$½^2 = ·25$
$2^3 = 8$	$½^3 = ·125$
$2^4 = 4^2 = 16$	$½^4 = ·0625$
$2^5 = 32$	$½^5 = ·03125$
$2^6 = 4^3 = 64$	$½^6 = ·015625$
$2^7 = 128$	$½^7 = ·0078125$
$2^8 = 4^4 = 256$	$½^8 = ·0039063$
$2^9 = 512$	$½^9 = ·0019531$
$2^{10} = 4^5 = 1024$	$½^{10} = ·00097656$

$3^2 = 9^1 = 9$	$1/3^2 = \cdot 11111$
$3^3 = 27$	$1/3^3 = \cdot 037037$
$3^4 = 9^2 = 81$	$1/3^4 = \cdot 012346$
$3^5 = 243$	$1/3^5 = \cdot 0041152$
$3^6 = 9^3 = 729$	$1/3^6 = \cdot 0013717$
$3^7 = 2187$	$1/3^7 = \cdot 00045721$
$3^8 = 9^4 = 6561$	$1/3^8 = \cdot 00015242$

$5^{\frac{1}{4}} = \sqrt{\sqrt{5}} =$	$1 \cdot 4954$	$1/5^{\frac{1}{4}} = \cdot 66872$
$5^{\frac{1}{2}} = \sqrt{5} =$	$2 \cdot 2361$	$1/5^{\frac{1}{2}} = \cdot 44721$
$5^{\frac{3}{4}} = \sqrt{\sqrt{5^3}} =$	$3 \cdot 3437$	$1/5^{\frac{3}{4}} = \cdot 29907$
$5^1 = 5 =$	$5 \cdot 0$	$1/5^1 = \cdot 2$
$5^{\frac{5}{4}} = 5\sqrt{\sqrt{5}} =$	$7 \cdot 4767$	$1/5^{\frac{5}{4}} = \cdot 13375$

$5^2 = 25$	$1/5^2 = \cdot 04$
$5^3 = 125$	$1/5^3 = \cdot 008$
$5^4 = 625$	$1/5^4 = \cdot 0016$
$5^5 = 3125$	$1/5^5 = \cdot 00032$
$5^6 = 15625$	$1/5^6 = \cdot 000064$

$$2^{\frac{1}{12}} = \sqrt[12]{2} = 1 \cdot 0595$$
$$2^{\frac{2}{12}} = 2^{\frac{1}{6}} = \sqrt[6]{2} = 1 \cdot 1225$$
$$2^{\frac{3}{12}} = 2^{\frac{1}{4}} = \sqrt[4]{2} = 1 \cdot 1892$$
$$2^{\frac{4}{12}} = 2^{\frac{1}{3}} = \sqrt[3]{2} = 1 \cdot 2599$$
$$2^{\frac{5}{12}} = \sqrt[12]{2^5} = 1 \cdot 3348$$
$$2^{\frac{6}{12}} = 2^{\frac{1}{2}} = \sqrt{2} = 1 \cdot 4142$$
$$2^{\frac{7}{12}} = \sqrt[12]{2^7} = 1 \cdot 4983$$
$$2^{\frac{8}{12}} = 2^{\frac{2}{3}} = \sqrt[3]{2^2} = 1 \cdot 5874$$
$$2^{\frac{9}{12}} = 2^{\frac{3}{4}} = \sqrt[4]{2^3} = 1 \cdot 6818$$
$$2^{\frac{10}{12}} = 2^{\frac{5}{6}} = \sqrt[6]{2^5} = 1 \cdot 7818$$
$$2^{\frac{11}{12}} = \sqrt[12]{2^{11}} = 1 \cdot 8877$$
$$2^{\frac{12}{12}} = 2^1 = 2 = 2 \cdot 0$$

$\sqrt{2} = 1 \cdot 4142$	$1/\sqrt{2} = \cdot 70711$
$\sqrt{3} = 1 \cdot 7321$	$1/\sqrt{3} = \cdot 57733$
$\sqrt{5} = 2 \cdot 2361$	$1/\sqrt{5} = \cdot 44721$
$\sqrt{6} = 2 \cdot 4495$	$1/\sqrt{6} = \cdot 40825$
$\sqrt{8} = 2 \cdot 8284$	$1/\sqrt{8} = \cdot 35356$
$\sqrt{10} = 3 \cdot 1623$	$1/\sqrt{10} = \cdot 31623$
$\sqrt{12} = 3 \cdot 4641$	$1/\sqrt{12} = \cdot 28868$
$\sqrt{15} = 3 \cdot 8730$	$1/\sqrt{15} = \cdot 25820$

RATIO TABLE

RELATING CENTS, SAVARTS, DECIMAL RATIOS AND MONOCHORD STRING-LENGTHS

In the first column are listed cents from 0 to 1200 in one cent intervals.

The second column gives the equivalent of the first column—in savarts.

The third column gives the equivalent of both the previous columns in decimal ratios.

The fourth column gives that string-length (in millimetres) whose note when sounded, produces, with the note sounded by 1000 millimetres of the same string subjected to the same physical conditions, that musical interval in the ear which corresponds with the ratio and logarithmic units given in the same row of the previous columns.

0	·0000	1·0000	1000·0	26	6·522	1·0151	985·1
1	·2509	1·0006	999·4	27	6·773	1·0157	984·5
2	·5017	1·0012	998·8	28	7·024	1·0163	984·0
3	·7526	1·0017	998·3	29	7·275	1·0169	983·4
4	1·003	1·0023	997·7	30	7·526	1·0175	982·8
5	1·254	1·0029	997·1	31	7·777	1·0181	982·2
6	1·505	1·0035	996·5	32	8·027	1·0187	981·6
7	1·756	1·0041	995·9	33	8·278	1·0192	981·2
8	2·007	1·0046	995·4	34	8·529	1·0198	980·6
9	2·258	1·0052	994·8	35	8·780	1·0204	980·0
10	2·509	1·0058	994·6	36	9·031	1·0210	979·4
11	2·759	1·0064	993·6	37	9·283	1·0216	978·9
12	3·010	1·0070	993·0	38	9·533	1·0222	978·3
13	3·261	1·0075	992·6	39	9·783	1·0228	977·7
14	3·512	1·0081	992·0	40	10·03	1·0234	977·1
15	3·763	1·0087	991·4	41	10·29	1·0240	976·6
16	4·014	1·0093	990·8	42	10·54	1·0246	976·0
17	4·265	1·0099	990·2	43	10·79	1·0252	975·4
18	4·515	1·0105	989·6	44	11·04	1·0257	974·9
19	4·766	1·0110	989·1	45	11·29	1·0263	974·4
20	5·017	1·0116	988·5	46	11·54	1·0269	973·8
21	5·268	1·0122	987·9	47	11·79	1·0275	973·2
22	5·519	1·0128	987·4	48	12·04	1·0281	972·7
23	5·770	1·0134	986·8	49	12·29	1·0287	972·1
24	6·021	1·0140	986·2	50	12·54	1·0293	971·5
25	6·271	1·0145	985·7	51	12·79	1·0299	971·0

52	13·04	1·0305	970·4
53	13·30	1·0311	969·8
54	13·55	1·0317	969·3
55	13·80	1·0323	968·7
56	14·05	1·0329	968·1
57	14·30	1·0335	967·6
58	14·55	1·0341	967·0
59	14·80	1·0347	966·5
60	15·05	1·0353	966·0
61	15·30	1·0359	965·3
62	15·55	1·0365	964·8
63	15·80	1·0371	964·2
64	16·05	1·0376	963·8
65	16·31	1·0383	963·1
66	16·57	1·0389	962·6
67	16·81	1·0395	962·0
68	17·06	1·0401	961·4
69	17·31	1·0407	960·9
70	17·56	1·0413	960·3
71	17·81	1·0419	959·8
72	18·06	1·0425	959·2
73	18·31	1·0431	958·7
74	18·56	1·0437	958·1
75	18·81	1·0443	957·6
76	19·07	1·0449	957·0
77	19·32	1·0455	956·5
78	19·57	1·0461	955·9
79	19·82	1·0467	955·4
80	20·07	1·0473	954·8
81	20·32	1·0479	954·3
82	20·57	1·0485	953·7
83	20·82	1·0491	953·2
84	21·07	1·0497	952·7
85	21·32	1·0503	952·1
86	21·57	1·0509	951·6
87	21·82	1·0515	951·0
88	22·08	1·0522	950·4
89	22·33	1·0528	949·8
90	22·58	1·0534	949·3
91	22·83	1·0540	948·8
92	23·08	1·0546	948·2
93	23·33	1·0552	947·7
94	23·58	1·0558	947·1
95	23·83	1·0564	946·6
96	24·08	1·0570	946·1
97	24·33	1·0576	945·5
98	24·58	1·0582	945·0
99	24·83	1·0588	944·5
100	25·09	1·0595	943·8
101	25·34	1·0601	943·3
102	25·59	1·0607	942·8
103	25·84	1·0613	942·2
104	26·09	1·0619	941·7
105	26·34	1·0625	941·2
106	26·59	1·0631	940·6
107	26·84	1·0638	940·0
108	27·09	1·0644	939·5
109	27·34	1·0650	939·0
110	27·59	1·0656	938·4
111	27·85	1·0662	937·9
112	28·10	1·0669	937·3
113	28·35	1·0675	936·8
114	28·60	1·0681	936·2
115	28·85	1·0687	935·7
116	29·10	1·0693	935·2
117	29·35	1·0699	934·7
118	29·60	1·0705	934·1
119	29·85	1·0712	933·5
120	30·10	1·0718	933·0
121	30·35	1·0724	932·5
122	30·60	1·0730	932·0
123	30·86	1·0736	931·4
124	31·11	1·0743	930·8
125	31·36	1·0749	930·3
126	31·61	1·0755	929·8
127	31·86	1·0761	929·3
128	32·11	1·0767	928·8
129	32·36	1·0774	928·2
130	32·61	1·0780	927·6
131	32·86	1·0786	927·1

132	33·11	1·0792	926·6
133	33·36	1·0798	926·1
134	33·62	1·0805	925·6
135	33·87	1·0811	925·0
136	34·12	1·0817	924·5
137	34·37	1·0824	923·9
138	34·62	1·0830	923·4
139	34·87	1·0836	922·8
140	35·12	1·0842	922·3
141	35·37	1·0849	921·7
142	35·62	1·0855	921·2
143	35·87	1·0861	920·7
144	36·12	1·0867	920·2
145	36·37	1·0874	919·6
146	36·63	1·0880	919·1
147	36·88	1·0886	918·6
148	37·13	1·0893	918·0
149	37·38	1·0899	917·5
150	37·63	1·0905	917·0
151	37·88	1·0911	916·5
152	38·13	1·0918	915·9
153	38·38	1·0924	915·4
154	38·63	1·0930	914·9
155	38·88	1·0937	914·3
156	39·13	1·0943	913·8
157	39·38	1·0949	913·3
158	39·64	1·0956	912·7
159	39·89	1·0962	912·2
160	40·14	1·0968	911·7
161	40·39	1·0975	911·0
162	40·64	1·0981	910·7
163	40·89	1·0987	910·2
164	41·14	1·0994	909·6
165	41·39	1·1000	909·0
166	41·64	1·1006	908·6
167	41·89	1·1013	908·0
168	42·14	1·1019	907·5
169	42·40	1·1026	906·9
170	42·65	1·1032	906·5
171	42·90	1·1038	906·0
172	43·15	1·1044	905·5
173	43·40	1·1051	904·9
174	43·65	1·1057	904·4
175	43·90	1·1064	903·8
176	44·15	1·1070	903·3
177	44·40	1·1076	902·9
178	44·65	1·1083	902·3
179	44·90	1·1089	901·8
180	45·15	1·1096	901·2
181	45·41	1·1102	900·7
182	45·66	1·1109	900·2
183	45·91	1·1115	899·7
184	46·16	1·1121	899·2
185	46·41	1·1128	898·6
186	46·66	1·1134	898·1
187	46·91	1·1141	897·6
188	47·16	1·1147	897·1
189	47·41	1·1153	896·6
190	47·66	1·1160	896·1
191	47·91	1·1166	895·6
192	48·16	1·1173	895·0
193	48·42	1·1180	894·6
194	48·67	1·1186	894·0
195	48·92	1·1192	893·5
196	49·17	1·1199	892·9
197	49·42	1·1205	892·5
198	49·67	1·1212	891·9
199	49·92	1·1218	891·4
200	50·17	1·1225	890·9
201	50·42	1·1231	890·5
202	50·67	1·1238	889·8
203	50·92	1·1244	889·4
204	51·18	1·1251	888·8
205	51·43	1·1257	888·3
206	51·68	1·1264	887·8
207	51·93	1·1270	887·3
208	52·18	1·1277	886·8
209	52·43	1·1283	886·3
210	52·68	1·1290	885·7
211	52·93	1·1296	885·3

212	53·18	1·1303	884·7
213	53·43	1·1309	884·2
214	53·68	1·1316	883·7
215	53·93	1·1322	883·2
216	54·19	1·1329	882·7
217	54·44	1·1335	882·2
218	54·69	1·1342	881·7
219	54·94	1·1349	881·1
220	55·19	1·1355	880·7
221	55·44	1·1362	880·1
222	55·69	1·1368	879·7
223	55·94	1·1375	879·1
224	56·19	1·1381	878·7
225	56·44	1·1388	878·1
226	56·69	1·1394	877·7
227	56·94	1·1401	877·1
228	57·20	1·1408	876·6
229	57·45	1·1414	876·1
230	57·70	1·1421	875·6
231	57·95	1·1427	875·1
232	58·20	1·1434	874·6
233	58·45	1·1441	874·0
234	58·70	1·1447	873·6
235	58·95	1·1454	873·1
236	59·20	1·1460	872·6
237	59·45	1·1467	872·1
238	59·70	1·1474	871·5
239	59·96	1·1480	871·1
240	60·21	1·1487	870·5
241	60·46	1·1494	870·0
242	60·71	1·1500	869·6
243	60·96	1·1507	869·0
244	61·21	1·1514	868·5
245	61·46	1·1520	868·1
246	61·71	1·1527	867·5
247	61·96	1·1533	867·1
248	62·21	1·1540	866·6
249	62·46	1·1547	866·0
250	62·71	1·1553	865·6
251	62·97	1·1560	865·1
252	63·22	1·1567	864·5
253	63·47	1·1574	864·0
254	63·72	1·1580	863·6
255	63·97	1·1587	863·0
256	64·22	1·1594	862·5
257	64·47	1·1600	862·1
258	64·72	1·1607	861·6
259	64·97	1·1614	861·0
260	65·22	1·1620	860·6
261	65·47	1·1627	860·1
262	65·72	1·1634	859·5
263	65·98	1·1641	859·0
264	66·23	1·1647	858·6
265	66·48	1·1654	858·1
266	66·73	1·1661	857·6
267	66·98	1·1668	857·0
268	67·23	1·1674	856·6
269	67·48	1·1681	856·1
270	67·73	1·1688	855·6
271	67·98	1·1695	855·0
272	68·23	1·1701	854·6
273	68·48	1·1708	854·1
274	68·74	1·1715	853·6
275	68·99	1·1722	853·1
276	69·24	1·1728	852·7
277	69·49	1·1735	852·2
278	69·74	1·1742	851·6
279	69·99	1·1749	851·1
280	70·24	1·1755	850·7
281	70·49	1·1762	850·2
282	70·74	1·1769	849·7
283	70·99	1·1776	849·2
284	71·24	1·1783	848·8
285	71·49	1·1789	848·2
286	71·75	1·1796	847·7
287	72·00	1·1804	847·2
288	72·25	1·1810	846·7
289	72·50	1·1817	846·2
290	72·75	1·1824	845·7
291	73·00	1·1831	845·2

292	73·25	1·1837	844·8
293	73·50	1·1844	844·3
294	73·75	1·1851	843·8
295	74·00	1·1858	843·3
296	74·25	1·1865	842·8
297	74·50	1·1871	842·4
298	74·76	1·1878	841·9
299	75·01	1·1885	841·4
300	75·26	1·1892	840·9
301	75·51	1·1899	840·4
302	75·76	1·1906	839·9
303	76·01	1·1913	839·4
304	76·26	1·1920	838·9
305	76·51	1·1926	838·5
306	76·76	1·1933	838·0
307	77·01	1·1940	837·5
308	77·26	1·1947	837·0
309	77·52	1·1954	836·5
310	77·77	1·1961	836·1
311	78·02	1·1968	835·6
312	78·27	1·1975	835·1
313	78·52	1·1982	834·6
314	78·77	1·1989	834·1
315	79·02	1·1996	833·6
316	79·27	1·2002	833·2
317	79·52	1·2009	832·7
318	79·77	1·2016	832·2
319	80·02	1·2023	831·7
320	80·27	1·2030	831·3
321	80·53	1·2037	830·8
322	80·78	1·2043	830·4
323	81·03	1·2051	829·8
324	81·28	1·2058	829·3
325	81·53	1·2065	828·8
326	81·78	1·2072	828·4
327	82·03	1·2079	827·9
328	82·28	1·2086	827·4
329	82·53	1·2093	826·9
330	82·78	1·2100	826·4
331	83·03	1·2107	826·0
332	83·28	1·2114	825·6
333	83·54	1·2121	825·0
334	83·79	1·2128	824·5
335	84·04	1·2135	824·1
336	84·29	1·2142	823·6
337	84·54	1·2149	823·1
338	84·79	1·2156	822·6
339	85·04	1·2163	822·0
340	85·29	1·2170	821·7
341	85·54	1·2177	821·3
342	85·79	1·2184	820·7
343	86·04	1·2191	820·3
344	86·30	1·2198	819·8
345	86·55	1·2205	819·3
346	86·80	1·2212	818·9
347	87·05	1·2219	818·4
348	87·30	1·2226	817·9
349	87·55	1·2234	817·4
350	87·80	1·2241	816·9
351	88·05	1·2248	816·5
352	88·30	1·2255	816·0
353	88·55	1·2262	815·5
354	88·80	1·2269	815·1
355	89·05	1·2276	814·7
356	89·31	1·2283	814·1
357	89·56	1·2290	813·7
358	89·81	1·2297	813·2
359	90·06	1·2304	812·7
360	90·31	1·2311	812·3
361	90·56	1·2319	811·8
362	90·81	1·2326	811·3
363	91·06	1·2333	811·0
364	91·31	1·2340	810·4
365	91·56	1·2347	809·9
366	91·81	1·2354	809·5
367	92·06	1·2361	809·0
368	92·32	1·2369	808·5
369	92·57	1·2376	808·1
370	92·82	1·2383	807·6
371	93·07	1·2390	807·1

372	93·32	1·2397	806·6	412	103·4	1·2688	788·1
373	93·57	1·2404	806·2	413	103·6	1·2694	787·8
374	93·82	1·2411	805·8	414	103·9	1·2703	787·3
375	94·07	1·2418	805·3	415	104·1	1·2709	786·8
376	94·32	1·2426	804·8	416	104·4	1·2717	786·3
377	94·57	1·2433	804·3	417	104·6	1·2723	786·0
378	94·82	1·2440	803·9	418	104·9	1·2732	785·4
379	95·08	1·2447	803·4	419	105·1	1·2738	785·0
380	95·33	1·2455	802·9	420	105·4	1·2747	784·5
381	95·58	1·2462	802·4	421	105·6	1·2753	784·1
382	95·83	1·2469	802·0	422	105·9	1·2762	783·6
383	96·08	1·2476	801·6	423	106·1	1·2767	783·3
384	96·33	1·2483	801·2	424	106·4	1·2776	782·7
385	96·58	1·2491	800·6	425	106·6	1·2782	782·4
386	96·83	1·2498	800·1	426	106·9	1·2791	781·8
387	97·08	1·2505	799·7	427	107·1	1·2797	781·4
388	97·33	1·2512	799·2	428	107·4	1·2806	780·9
389	97·58	1·2519	798·8	429	107·6	1·2811	780·6
390	97·83	1·2527	798·3	430	107·9	1·2820	780·0
391	98·09	1·2534	797·8	431	108·1	1·2826	779·7
392	98·34	1·2541	797·4	432	108·4	1·2835	779·1
393	98·59	1·2549	796·9	433	108·6	1·2841	778·6
394	98·84	1·2556	796·4	434	108·9	1·2850	778·2
395	99·09	1·2563	796·0	435	109·1	1·2856	777·8
396	99·34	1·2570	795·5	436	109·4	1·2865	777·3
397	99·59	1·2577	795·1	437	109·6	1·2871	776·9
398	99·84	1·2585	794·6	438	109·9	1·2880	776·4
399	100·1	1·2592	794·2	439	110·1	1·2885	776·1
400	100·3	1·2598	793·8	440	110·4	1·2894	775·6
401	100·6	1·2607	793·2	441	110·6	1·2900	775·2
402	100·8	1·2613	792·8	442	110·9	1·2909	774·7
403	101·1	1·2621	792·3	443	111·1	1·2915	774·3
404	101·3	1·2627	792·0	444	111·4	1·2924	773·8
405	101·6	1·2636	791·4	445	111·6	1·2930	773·4
406	101·8	1·2642	791·0	446	111·9	1·2939	772·9
407	102·1	1·2650	790·5	447	112·1	1·2945	772·5
408	102·4	1·2659	790·0	448	112·4	1·2954	772·0
409	102·6	1·2665	789·6	449	112·6	1·2960	771·6
410	102·9	1·2674	789·0	450	112·9	1·2969	771·1
411	103·1	1·2680	788·6	451	113·1	1·2975	770·7

452	113·4	1·2984	770·2
453	113·6	1·2990	769·8
454	113·9	1·2999	769·3
455	114·1	1·3005	768·9
456	114·4	1·3014	768·4
457	114·6	1·3020	768·0
458	114·9	1·3029	767·5
459	115·1	1·3035	767·2
460	115·4	1·3044	766·6
461	115·6	1·3050	766·3
462	115·9	1·3059	765·8
463	116·1	1·3065	765·4
464	116·4	1·3074	764·9
465	116·6	1·3080	764·5
466	116·9	1·3089	764·0
467	117·2	1·3098	763·5
468	117·4	1·3104	763·1
469	117·7	1·3113	762·6
470	117·9	1·3119	762·3
471	118·2	1·3128	761·7
472	118·4	1·3134	761·4
473	118·7	1·3143	760·9
474	118·9	1·3149	760·5
475	119·2	1·3158	760·0
476	119·4	1·3164	759·6
477	119·7	1·3173	759·1
478	119·9	1·3180	758·7
479	120·2	1·3189	758·2
480	120·4	1·3195	757·9
481	120·7	1·3204	757·3
482	120·9	1·3210	757·0
483	121·2	1·3219	756·5
484	121·4	1·3225	756·1
485	121·7	1·3234	755·6
486	121·9	1·3241	755·2
487	122·2	1·3250	754·7
488	122·4	1·3256	754·4
489	122·7	1·3265	753·9
490	122·9	1·3271	753·5
491	123·2	1·3280	753·0
492	123·4	1·3286	752·7
493	123·7	1·3295	752·2
494	123·9	1·3301	751·8
495	124·2	1·3311	751·3
496	124·4	1·3317	750·9
497	124·7	1·3326	750·5
498	124·9	1·3332	750·1
499	125·2	1·3341	749·6
500	125·4	1·3348	749·2
501	125·7	1·3357	748·7
502	125·9	1·3363	748·3
503	126·2	1·3372	747·8
504	126·4	1·3378	747·5
505	126·7	1·3388	746·9
506	126·9	1·3394	746·6
507	127·2	1·3403	746·1
508	127·4	1·3409	745·8
509	127·7	1·3418	745·3
510	127·9	1·3424	744·9
511	128·2	1·3434	744·4
512	128·4	1·3440	744·0
513	128·7	1·3449	743·5
514	128·9	1·3456	743·2
515	129·2	1·3465	742·7
516	129·4	1·3471	742·3
517	129·7	1·3480	741·8
518	129·9	1·3487	741·5
519	130·2	1·3496	741·0
520	130·4	1·3502	740·6
521	130·7	1·3511	740·1
522	130·9	1·3518	739·8
523	131·2	1·3527	739·3
524	131·4	1·3533	738·9
525	131·7	1·3543	738·4
526	132·0	1·3552	737·9
527	132·2	1·3558	737·6
528	132·5	1·3568	737·0
529	132·7	1·3574	736·7
530	133·0	1·3583	736·2
531	133·2	1·3589	735·9

532	133·5	1·3599	735·3
533	133·7	1·3605	735·0
534	134·0	1·3614	734·5
535	134·2	1·3621	734·2
536	134·5	1·3630	733·7
537	134·7	1·3636	733·3
538	135·0	1·3646	732·8
539	135·2	1·3652	732·5
540	135·5	1·3662	732·0
541	135·7	1·3668	731·6
542	136·0	1·3677	731·2
543	136·2	1·3684	730·8
544	136·5	1·3693	730·3
545	136·7	1·3699	730·0
546	137·0	1·3709	729·4
547	137·2	1·3715	729·1
548	137·5	1·3725	728·6
549	137·7	1·3731	728·3
550	138·0	1·3740	727·8
551	138·2	1·3747	727·4
552	138·5	1·3756	727·0
553	138·7	1·3763	726·6
554	139·0	1·3772	726·1
555	139·2	1·3778	725·8
556	139·5	1·3788	725·3
557	139·7	1·3794	725·0
558	140·0	1·3804	724·4
559	140·2	1·3810	724·1
560	140·5	1·3820	723·6
561	140·7	1·3826	723·3
562	141·0	1·3836	722·8
563	141·2	1·3842	722·4
564	141·5	1·3852	721·9
565	141·7	1·3858	721·6
566	142·0	1·3868	721·1
567	142·2	1·3874	720·8
568	142·5	1·3884	720·2
569	142·7	1·3890	719·9
570	143·0	1·3900	719·4
571	143·2	1·3906	719·2
572	143·5	1·3916	718·6
573	143·7	1·3922	718·3
574	144·0	1·3932	717·8
575	144·2	1·3938	717·5
576	144·5	1·3948	716·9
577	144·7	1·3954	716·6
578	145·0	1·3964	716·1
579	145·2	1·3970	715·8
580	145·5	1·3980	715·3
581	145·7	1·3986	715·0
582	146·0	1·3996	714·5
583	146·3	1·4006	714·0
584	146·6	1·4015	713·5
585	146·8	1·4022	713·2
586	147·0	1·4029	712·8
587	147·3	1·4038	712·4
588	147·6	1·4048	711·8
589	147·8	1·4054	711·5
590	148·0	1·4060	711·2
591	148·3	1·4070	710·7
592	148·6	1·4080	710·2
593	148·8	1·4086	709·9
594	149·0	1·4093	709·6
595	149·3	1·4103	709·0
596	149·5	1·4109	708·8
597	149·8	1·4119	708·3
598	150·0	1·4125	708·0
599	150·3	1·4135	707·5
600	150·5	1·4142	707·1
601	150·8	1·4151	706·7
602	151·0	1·4158	706·3
603	151·3	1·4168	705·8
604	151·5	1·4174	705·5
605	151·8	1·4184	705·0
606	152·0	1·4190	704·7
607	152·3	1·4200	704·2
608	152·5	1·4207	703·9
609	152·8	1·4217	703·4
610	153·0	1·4223	703·1
611	153·3	1·4233	702·6

612	153·5	1·4240	702·2	652	163·6	1·4575	686·1
613	153·8	1·4250	701·8	653	163·8	1·4581	685·8
614	154·0	1·4256	701·5	654	164·1	1·4592	685·3
615	154·3	1·4266	701·0	655	164·3	1·4598	685·0
616	154·5	1·4272	700·7	656	164·6	1·4608	684·6
617	154·8	1·4282	700·1	657	164·8	1·4615	684·2
618	155·0	1·4289	699·8	658	165·1	1·4625	683·8
619	155·3	1·4299	699·3	659	165·3	1·4632	683·4
620	155·5	1·4305	699·1	660	165·6	1·4642	683·0
621	155·8	1·4315	698·6	661	165·8	1·4649	682·6
622	156·0	1·4322	698·2	662	166·1	1·4659	682·2
623	156·3	1·4332	697·7	663	166·3	1·4666	681·8
624	156·5	1·4338	697·4	664	166·6	1·4676	681·4
625	156·8	1·4348	697·0	665	166·8	1·4683	681·1
626	157·0	1·4355	696·6	666	167·1	1·4693	680·6
627	157·3	1·4365	696·1	667	167·3	1·4699	680·3
628	157·5	1·4371	695·8	668	167·6	1·4710	679·8
629	157·8	1·4381	695·4	669	167·8	1·4716	679·5
630	158·0	1·4388	695·0	670	168·1	1·4727	679·0
631	158·3	1·4398	694·5	671	168·3	1·4733	678·7
632	158·5	1·4405	694·2	672	168·6	1·4743	678·3
633	158·8	1·4415	693·7	673	168·8	1·4750	678·0
634	159·0	1·4421	693·4	674	169·1	1·4760	677·5
635	159·3	1·4431	693·0	675	169·3	1·4767	677·2
636	159·5	1·4438	692·6	676	169·6	1·4777	676·7
637	159·8	1·4448	692·1	677	169·8	1·4784	676·4
638	160·0	1·4454	691·9	678	170·1	1·4794	675·9
639	160·3	1·4464	691·4	679	170·3	1·4801	675·6
640	160·5	1·4471	691·0	680	170·6	1·4812	675·1
641	160·8	1·4481	690·6	681	170·8	1·4818	674·9
642	161·1	1·4491	690·1	682	171·1	1·4829	674·4
643	161·3	1·4498	689·8	683	171·3	1·4835	674·1
644	161·6	1·4508	689·3	684	171·6	1·4846	673·6
645	161·8	1·4514	689·0	685	171·8	1·4853	673·3
646	162·1	1·4524	688·5	686	172·1	1·4863	672·8
647	162·3	1·4531	688·2	687	172·3	1·4870	672·5
648	162·6	1·4541	687·7	688	172·6	1·4880	672·0
649	162·8	1·4548	687·4	689	172·8	1·4887	671·7
650	163·0	1·4555	687·0	690	173·1	1·4897	671·3
651	163·3	1·4565	686·6	691	173·3	1·4904	671·0

692	173·6	1·4914	670·5
693	173·8	1·4921	670·2
694	174·1	1·4931	669·8
695	174·3	1·4938	669·4
696	174·6	1·4949	668·9
697	174·8	1·4955	668·7
698	175·1	1·4966	668·2
699	175·3	1·4973	667·9
700	175·6	1·4983	667·4
701	175·9	1·4993	667·0
702	176·1	1·5000	666·7
703	176·4	1·5011	666·2
704	176·6	1·5018	665·9
705	176·9	1·5028	665·4
706	177·1	1·5035	665·1
707	177·4	1·5045	664·7
708	177·6	1·5052	664·4
709	177·9	1·5063	663·9
710	178·1	1·5070	663·6
711	178·4	1·5080	663·1
712	178·6	1·5087	662·8
713	178·9	1·5097	662·4
714	179·1	1·5104	662·1
715	179·4	1·5115	661·6
716	179·6	1·5122	661·3
717	179·9	1·5132	660·9
718	180·1	1·5139	660·5
719	180·4	1·5150	660·1
720	180·6	1·5157	659·8
721	180·9	1·5167	659·3
722	181·1	1·5174	659·0
723	181·4	1·5184	658·6
724	181·6	1·5191	658·3
725	181·9	1·5202	657·8
726	182·1	1·5209	657·5
727	182·4	1·5219	657·1
728	182·6	1·5227	656·7
729	182·9	1·5237	656·3
730	183·1	1·5244	656·0
731	183·4	1·5255	655·5
732	183·6	1·5262	655·2
733	183·9	1·5272	654·8
734	184·1	1·5279	654·5
735	184·4	1·5290	654·0
736	184·6	1·5297	653·7
737	184·9	1·5307	653·3
738	185·1	1·5314	653·0
739	185·4	1·5325	652·5
740	185·6	1·5332	652·2
741	185·9	1·5343	651·8
742	186·1	1·5350	651·5
743	186·4	1·5360	651·0
744	186·6	1·5367	650·7
745	186·9	1·5378	650·3
746	187·1	1·5385	650·0
747	187·4	1·5396	649·5
748	187·6	1·5403	649·2
749	187·9	1·5413	648·8
750	188·1	1·5421	648·0
751	188·4	1·5431	648·0
752	188·6	1·5438	647·6
753	188·9	1·5449	647·3
754	189·1	1·5456	647·0
755	189·4	1·5467	646·5
756	189·6	1·5474	646·2
757	189·9	1·5485	645·9
758	190·2	1·5495	645·4
759	190·4	1·5502	645·1
760	190·7	1·5513	644·6
761	190·9	1·5520	644·3
762	191·2	1·5531	643·9
763	191·4	1·5538	643·6
764	191·7	1·5549	643·1
765	191·9	1·5556	642·8
766	192·2	1·5567	642·4
767	192·4	1·5574	642·1
768	192·7	1·5585	641·6
769	192·9	1·5592	641·4
770	193·2	1·5603	640·9
771	193·4	1·5610	640·6

772	193·7	1·5621	640·2	812	203·7	1·5985	625·6
773	193·9	1·5628	639·9	813	203·9	1·5992	625·3
774	194·2	1·5639	639·4	814	204·2	1·6003	624·9
775	194·4	1·5646	639·1	815	204·4	1·6010	624·6
776	194·7	1·5657	638·7	816	204·7	1·6021	624·2
777	194·9	1·5664	638·4	817	205·0	1·6032	623·8
778	195·2	1·5675	638·0	818	205·2	1·6040	623·4
779	195·4	1·5682	637·7	819	205·5	1·6051	623·0
780	195·7	1·5693	637·2	820	205·7	1·6058	622·7
781	195·9	1·5700	636·9	821	206·0	1·6070	622·3
782	196·2	1·5711	636·5	822	206·2	1·6077	622·0
783	196·4	1·5718	636·2	823	206·5	1·6088	621·6
784	196·7	1·5729	635·8	824	206·7	1·6095	621·3
785	196·9	1·5736	635·5	825	207·0	1·6106	620·9
786	197·2	1·5747	635·0	826	207·2	1·6114	620·6
787	197·4	1·5754	634·8	827	207·5	1·6125	620·2
788	197·7	1·5765	634·3	828	207·7	1·6132	619·9
789	197·9	1·5772	634·0	829	208·0	1·6144	619·4
790	198·2	1·5783	633·6	830	208·2	1·6151	619·2
791	198·4	1·5791	633·3	831	208·5	1·6162	618·7
792	198·7	1·5802	632·8	832	208·7	1·6170	618·4
793	198·9	1·5809	632·6	833	209·0	1·6181	618·0
794	199·2	1·5820	632·1	834	209·2	1·6188	617·7
795	199·4	1·5827	631·8	835	209·5	1·6200	617·3
796	199·7	1·5838	631·4	836	209·7	1·6207	617·0
797	199·9	1·5845	631·1	837	210·0	1·6219	616·6
798	200·2	1·5856	630·7	838	210·2	1·6226	616·3
799	200·4	1·5864	630·4	839	210·5	1·6237	615·9
800	200·7	1·5874	630·0	840	210·7	1·6244	615·6
801	200·9	1·5882	629·6	841	211·0	1·6255	615·2
802	201·2	1·5893	629·2	842	211·2	1·6263	614·9
803	201·4	1·5900	628·9	843	211·5	1·6274	614·5
804	201·7	1·5911	628·5	844	211·7	1·6282	614·2
805	201·9	1·5918	628·2	845	212·0	1·6293	613·8
806	202·2	1·5930	627·7	846	212·2	1·6300	613·5
807	202·4	1·5937	627·5	847	212·5	1·6312	613·0
808	202·7	1·5948	627·0	848	212·7	1·6319	612·8
809	202·9	1·5955	626·8	849	213·0	1·6331	612·3
810	203·2	1·5966	626·3	850	213·2	1·6338	612·1
811	203·4	1·5973	626·1	851	213·5	1·6349	611·7

852	213·7	1·6357	611·4	892	223·8	1·6742	597·3
853	214·0	1·6369	610·9	893	224·0	1·6750	597·0
854	214·2	1·6376	610·6	894	224·3	1·6761	596·6
855	214·5	1·6387	610·2	895	224·5	1·6769	596·3
856	214·7	1·6395	609·9	896	224·8	1·6780	595·9
857	215·0	1·6406	609·5	897	225·0	1·6788	595·7
858	215·2	1·6413	609·3	898	225·3	1·6800	595·2
859	215·5	1·6425	608·8	899	225·5	1·6807	595·0
860	215·7	1·6432	608·6	900	225·8	1·6819	594·6
861	216·0	1·6444	608·1	901	226·0	1·6827	594·3
862	216·2	1·6451	607·9	902	226·3	1·6838	593·9
863	216·5	1·6463	607·4	903	226·5	1·6846	593·6
864	216·7	1·6470	607·2	904	226·8	1·6858	593·2
865	217·0	1·6481	606·8	905	227·0	1·6866	592·9
866	217·2	1·6489	606·5	906	227·3	1·6877	592·5
867	217·5	1·6501	606·0	907	227·5	1·6885	592·2
868	217·7	1·6508	605·8	908	227·8	1·6897	591·8
869	218·0	1·6520	605·3	909	228·0	1·6905	591·5
870	218·2	1·6527	605·1	910	228·3	1·6916	591·2
871	218·5	1·6539	604·6	911	228·5	1·6924	590·9
872	218·7	1·6546	604·4	912	228·8	1·6935	590·5
873	219·0	1·6558	603·9	913	229·0	1·6943	590·2
874	219·3	1·6569	603·5	914	229·3	1·6955	589·8
875	219·5	1·6577	603·2	915	229·5	1·6963	589·5
876	219·8	1·6588	602·8	916	229·8	1·6975	589·1
877	220·0	1·6596	602·6	917	230·0	1·6982	588·9
878	220·3	1·6607	602·2	918	230·3	1·6994	588·4
879	220·5	1·6615	601·9	919	230·5	1·7005	588·1
880	220·8	1·6626	601·5	920	230·8	1·7014	587·8
881	221·0	1·6634	601·2	921	231·0	1·7022	587·5
882	221·3	1·6646	600·7	922	231·3	1·7033	587·1
883	221·5	1·6653	600·5	923	231·5	1·7041	586·8
884	221·8	1·6665	600·0	924	231·8	1·7053	586·4
885	220·0	1·6672	599·8	925	232·0	1·7061	586·1
886	222·3	1·6684	599·4	926	232·3	1·7073	585·7
887	222·5	1·6692	599·1	927	232·5	1·7080	585·5
888	222·8	1·6703	598·7	928	232·8	1·7092	585·1
889	223·0	1·6711	598·4	929	233·0	1·7101	584·8
890	223·3	1·6722	598·0	930	233·3	1·7112	584·4
891	223·5	1·6730	597·7	931	233·5	1·7120	584·1

932	233·8	1·7132	583·7	972	243·8	1·7531	570·4
933	234·1	1·7144	583·3	973	244·1	1·7543	570·0
934	234·3	1·7151	583·0	974	244·3	1·7551	569·8
935	234·6	1·7163	582·6	975	244·6	1·7563	569·4
936	234·8	1·7171	582·4	976	244·8	1·7571	569·1
937	235·1	1·7183	582·0	977	245·1	1·7583	568·7
938	235·3	1·7191	581·7	978	245·3	1·7591	568·5
939	235·5	1·7199	581·4	979	245·6	1·7604	568·1
940	235·8	1·7211	581·0	980	245·8	1·7612	567·8
941	236·1	1·7223	580·6	981	246·1	1·7624	567·4
942	236·3	1·7231	580·3	982	246·3	1·7632	567·2
943	236·6	1·7242	580·0	983	246·6	1·7644	566·8
944	236·8	1·7250	579·7	984	246·8	1·7652	566·5
945	237·1	1·7262	579·3	985	247·1	1·7664	566·1
946	237·3	1·7270	579·0	986	247·3	1·7673	565·8
947	237·6	1·7282	578·6	987	247·6	1·7685	565·6
948	237·8	1·7290	578·2	988	247·8	1·7693	565·2
949	238·1	1·7302	578·0	989	248·1	1·7705	564·8
950	238·3	1·7310	577·7	990	248·3	1·7713	564·6
951	238·6	1·7322	577·3	991	248·6	1·7726	564·1
952	238·8	1·7330	576·9	992	248·9	1·7738	563·8
953	239·1	1·7342	576·6	993	249·1	1·7746	563·5
954	239·3	1·7350	576·4	994	249·3	1·7754	563·3
955	239·6	1·7362	576·0	995	249·6	1·7766	582·9
956	239·8	1·7370	575·7	996	249·9	1·7779	562·5
957	240·1	1·7382	575·3	997	250·1	1·7787	562·2
958	240·3	1·7390	575·0	998	250·4	1·7799	561·8
959	240·6	1·7402	574·6	999	250·6	1·7807	561·6
960	240·8	1·7410	574·4	1000	250·9	1·7820	561·2
961	241·1	1·7422	574·0	1001	251·1	1·7828	560·9
962	241·3	1·7430	573·7	1002	251·4	1·7840	560·5
963	241·6	1·7442	573·3	1003	251·6	1·7848	560·3
964	241·8	1·7450	573·1	1004	251·9	1·7861	559·9
965	242·1	1·7462	572·7	1005	252·1	1·7869	559·6
966	242·3	1·7470	572·4	1006	252·4	1·7881	559·3
967	242·6	1·7482	572·0	1007	252·6	1·7890	559·0
968	242·8	1·7490	571·8	1008	252·9	1·7902	558·6
969	243·1	1·7502	571·4	1009	253·1	1·7910	558·3
970	243·3	1·7511	571·1	1010	253·4	1·7923	558·0
971	243·6	1·7523	570·7	1011	253·6	1·7931	557·7

1012	253·9	1·7943	557·3
1013	254·1	1·7951	557·1
1014	254·4	1·7964	556·7
1015	254·6	1·7972	556·4
1016	254·9	1·7984	556·0
1017	255·1	1·7993	555·8
1018	255·4	1·8005	555·4
1019	255·6	1·8014	555·1
1020	255·9	1·8026	554·8
1021	256·1	1·8034	554·5
1022	256·4	1·8047	554·1
1023	256·6	1·8055	553·9
1024	256·9	1·8068	553·5
1025	257·1	1·8076	553·2
1026	257·4	1·8088	552·9
1027	257·6	1·8097	552·6
1028	257·9	1·8109	552·2
1029	258·1	1·8118	551·9
1030	258·4	1·8130	551·6
1031	258·6	1·8138	551·3
1032	258·9	1·8151	550·9
1033	259·1	1·8159	550·7
1034	259·4	1·8172	550·3
1035	259·6	1·8180	550·1
1036	259·9	1·8193	549·7
1037	260·1	1·8201	549·4
1038	260·4	1·8214	549·0
1039	260·6	1·8222	548·8
1040	260·9	1·8235	548·4
1041	261·1	1·8243	548·2
1042	261·4	1·8256	547·8
1043	261·6	1·8264	547·5
1044	261·9	1·8277	547·1
1045	262·1	1·8285	546·9
1046	262·4	1·8298	546·5
1047	262·6	1·8306	546·3
1048	262·9	1·8319	545·9
1049	263·2	1·8332	545·5
1050	263·4	1·8340	545·3
1051	263·7	1·8353	544·9
1052	263·9	1·8361	544·6
1053	264·2	1·8374	544·2
1054	264·4	1·8382	544·0
1055	264·7	1·8395	543·6
1056	264·9	1·8404	543·4
1057	265·2	1·8416	543·0
1058	265·4	1·8425	542·7
1059	265·7	1·8437	542·4
1060	265·9	1·8446	542·1
1061	266·2	1·8459	541·7
1062	266·4	1·8467	541·5
1063	266·7	1·8480	541·1
1064	266·9	1·8488	540·9
1065	267·2	1·8501	540·5
1066	267·4	1·8510	540·2
1067	267·7	1·8523	539·9
1068	267·9	1·8531	539·6
1069	268·2	1·8544	539·3
1070	268·4	1·8552	539·0
1071	268·7	1·8565	538·6
1072	268·9	1·8574	538·4
1073	269·2	1·8587	538·0
1074	269·4	1·8595	537·8
1075	269·7	1·8608	537·4
1076	269·9	1·8617	537·1
1077	270·2	1·8629	536·8
1078	270·4	1·8638	536·5
1079	270·7	1·8651	536·2
1080	270·9	1·8660	535·9
1081	271·2	1·8672	535·6
1082	271·4	1·8681	535·3
1083	271·7	1·8694	534·9
1084	271·9	1·8703	534·7
1085	272·2	1·8715	534·3
1086	272·4	1·8724	534·1
1087	272·7	1·8737	533·7
1088	272·9	1·8746	533·4
1089	273·2	1·8759	533·1
1090	273·4	1·8767	532·9
1091	273·7	1·8780	532·5

1092	273·9	1·8789	532·2
1093	274·2	1·8802	531·9
1094	274·4	1·8810	531·6
1095	274·7	1·8823	531·3
1096	274·9	1·8832	531·0
1097	275·2	1·8845	530·6
1098	275·4	1·8854	530·4
1099	275·7	1·8867	530·0
1100	275·9	1·8876	529·8
1101	276·2	1·8889	529·4
1102	276·4	1·8897	529·2
1103	276·7	1·8910	528·8
1104	276·9	1·8919	528·6
1105	277·2	1·8932	528·2
1106	277·4	1·8941	528·0
1107	277·7	1·8954	527·6
1108	278·0	1·8968	527·2
1109	278·2	1·8976	527·0
1110	278·5	1·8989	526·6
1111	278·7	1·8998	526·4
1112	279·0	1·9011	526·0
1113	279·2	1·9020	525·8
1114	279·5	1·9033	525·4
1115	279·7	1·9041	525·2
1116	280·0	1·9055	524·8
1117	280·2	1·9063	524·6
1118	280·5	1·9077	524·2
1119	280·7	1·9085	524·0
1120	281·0	1·9099	523·6
1121	281·2	1·9107	523·4
1122	281·5	1·9121	523·0
1123	281·7	1·9129	522·8
1124	282·0	1·9143	522·4
1125	282·2	1·9151	522·2
1126	282·5	1·9165	521·8
1127	282·7	1·9173	521·6
1128	283·0	1·9187	521·2
1129	283·2	1·9196	529·9
1130	283·5	1·9209	520·6
1131	283·7	1·9218	520·3
1132	284·0	1·9231	520·0
1133	284·2	1·9240	519·6
1134	284·5	1·9253	519·4
1135	284·7	1·9262	519·2
1136	285·0	1·9275	518·8
1137	285·2	1·9284	518·6
1138	285·5	1·9297	518·2
1139	285·7	1·9306	518·0
1140	286·0	1·9320	517·6
1141	286·2	1·9329	517·4
1142	286·5	1·9342	517·0
1143	286·7	1·9351	516·8
1144	287·0	1·9364	516·4
1145	287·2	1·9373	516·2
1146	287·5	1·9387	515·8
1147	287·7	1·9395	515·6
1148	288·0	1·9409	515·2
1149	288·2	1·9418	515·0
1150	288·5	1·9431	514·6
1151	288·7	1·9440	514·4
1152	289·0	1·9454	514·0
1153	289·2	1·9463	513·8
1154	289·5	1·9476	513·5
1155	289·7	1·9485	513·2
1156	290·0	1·9499	512·8
1157	290·2	1·9507	512·6
1158	290·5	1·9521	512·3
1159	290·7	1·9530	512·0
1160	291·0	1·9543	511·7
1161	291·2	1·9552	511·5
1162	291·5	1·9566	511·1
1163	291·7	1·9575	510·9
1164	292·0	1·9589	510·5
1165	292·2	1·9597	510·3
1166	292·5	1·9611	509·9
1167	292·8	1·9625	509·6
1168	293·0	1·9634	509·3
1169	293·3	1·9647	509·0
1170	293·5	1·9656	508·8
1171	293·8	1·9670	508·4

1172	294·0	1·9679	508·2
1173	294·3	1·9692	507·8
1174	294·5	1·9702	507·6
1175	294·8	1·9715	507·2
1176	295·0	1·9724	507·0
1177	295·3	1·9738	506·6
1178	295·5	1·9747	506·4
1179	295·8	1·9761	506·1
1180	296·0	1·9770	505·8
1181	296·3	1·9783	505·5
1182	296·5	1·9792	505·3
1183	296·8	1·9806	504·9
1184	297·0	1·9815	504·7
1185	297·3	1·9829	504·3
1186	297·5	1·9838	504·1
1187	297·8	1·9852	503·7
1188	298·0	1·9861	503·5
1189	298·3	1·9875	503·1
1190	298·5	1·9884	502·9
1191	298·8	1·9898	502·6
1192	299·0	1·9907	502·3
1193	299·3	1·9920	502·0
1194	299·5	1·9930	501·8
1195	299·8	1·9943	501·4
1196	300·0	1·9953	501·2
1197	300·3	1·9966	500·9
1198	300·5	1·9976	500·6
1199	300·8	1·9989	500·3
1200	301·0	2·0000	500·0

INDEX

Those items already listed under chapter headings are not necessarily repeated in this index.

Page numbers in **heavy type** indicate definitions.